AF334011

Probabilities on Algebraic Structures

ULF GRENANDER

L. Herbert Ballou University Professor Emeritus
BROWN UNIVERSITY

Dover Publications, Inc.
Mineola, New York

Bibliographical Note

This Dover edition, first published in 2008, is an unabridged republication of the work originally published by Almqvist & Wiksell, Stockholm, and John Wiley & Sons, Inc., New York, in 1963.

Library of Congress Cataloging-in-Publication Data

Grenander, Ulf.
    Probabilities on algebraic structures. — Dover ed. / Ulf Grenander
        p. cm.
    "This Dover edition, first published in 2008, is an unabridged republication of the work originally published by Almqvist & Wiksell, Stockholm, and John Wiley & Sons, Inc., New York, in 1963."
    ISBN-13: 978-0-486-46287-5
    ISBN-10: 0-486-46287-0
    1. Probablilities. 2. Algebraic topology. 3. Functional analysis. I. Title.

QA273.G69 2008
519.2—dc22

                                                         2007035600

Manufactured in the United States of America
Dover Publications, Inc., 31 East 2nd Street, Mineola, N.Y. 11501

To my parents

INTRODUCTION

This book attempts to present a unified and coherent theory of the calculus of probability on algebraic structures. Among such structures we shall consider some that merit special attention because of the role they play in the applications of the theory as well as because of their intrinsic interest. These are topological semi-groups and groups, topological vector spaces and algebras.

The author became interested in the subject through a number of practical problems, some of which are described in chapter 1. Therefore the emphasis will be on concrete results that can actually be used and that lead, at least in principle, to solutions that can be determined exactly or approximately, analytically or numerically. We do not try to reach results of the most general character or to develop the theory to its most refined form. Therefore we have not hesitated to impose conditions like separability, Borel measurability etc. in situations where this may not be necessary but where such conditions result in simplifications. Some reader may feel inclined to remove such imperfections, to reach necessary and sufficient conditions and so on. The author hopes that this will be done in the future.

To avoid obscuring the main ideas of the theory by lengthy computations and technical arguments such manipulations have sometimes been sketched only. This has been done especially when references could be cited to the literature where a complete treatment has been given.

To make the presentation as concrete as possible we will discuss a number of special cases at the end of every chapter. The author believes that some of these cases are as important as the theorems that they exemplify. They may sometimes indicate how to extend the theory.

The reader should observe that all the references as well as supplementary information are given in the Notes at the end of the book. Chapter 1 contains an outline of the history of the subject and also some remarks concerning the practical background to the theory. The arguments in this chapter are of a heuristic nature and will reappear in a rigorous form in the later chapters.

At a first reading the reader may prefer to skip sections 4.2 (where the very technical proofs are only sketched), 5.4 and 6.6 (which are of special nature).

It is difficult to specify the necessary prerequisites for reading the book, but it seems clear that the standard "calculus and mathematical maturity" would scarcely be adequate. Since the book is written for probabilists it will be assumed that the reader is well acquainted with probability and measure theory, say corresponding to the content of Loève: *Probability Theory* and Halmos: *Measure Theory*. Many of the measurability arguments used are of standard type and will only be hinted at. The reader should also have some knowledge of basic topological algebra. Neumark: *Normierte Algebren* is warmly recommended as a lucid and up to date presentation. Since this topic may be less well known among probabilists it seemed appropriate to be more complete when discussing related questions and the reader will find a number of the fundamental definitions and logical relations in the Notes. We must also ask of the reader that he know the elements of functional analysis. A suitable book would be Hille-Phillips: *Functional Analysis and Semi-groups*, which also contains some highly relevant information about semi-groups; this is almost indispensable when studying homogeneous processes.

The object of our study is the probability distribution on a structure in which at least one binary algebraic operation has been defined in such a way that it is continuous in a suitable topology. We can then talk of a stochastic element drawn at random from the structure. Composing two such independent stochastic elements via a binary operation we are led to convolutions (this is the key word!) of two probability distributions. To a considerable extent we shall study how convolutions behave, especially when we perform many of them. Our approach will be analytical, the main tool being Fourier analysis. There is no doubt that this is the correct approach if we want definite results, algorithms etc. There is however another way, more algebraic or perhaps probabilistic, in manner. We then consider the set of probability distributions in question as forming a topological semi-group. Applying semi-group theory we can reach certain results of considerable interest. Actually this is quite an attractive application of the general theory of semi-groups, but it will not be discussed more than occasionally in the text.

I have had a number of valuable discussions with colleagues whom I would like to thank for their suggestions: R. Fortet, E. Mourier, L. Schmetterer, Z. Šidak, and A. Špaček. I am very grateful to D. Wehn for putting a manuscript at my disposal before its publication. Sections 4.2–4.4. lean heavily on Wehn's results and exposition. M. Rosenblatt and W. Freibeiger read the entire manuscript and suggested many changes. R. Loynes and P. Martin-Löf scrutinized the book in detail and I am very grateful to them for their work. They spotted a number of mistakes or obscurities and also contributed some essential results to the theory.

I would also like to thank Försäkringsaktiebolaget Skandia and Statens Naturvetenskapliga Forskningsråd (Swedish Natural Science Research Council) for their financial support.

ULF GRENANDER

CONTENTS

CHAPTER 1

—

HISTORICAL BACKGROUND AND
PRACTICAL MOTIVATION OF THE PROBLEM

1.1. Why study probabilities on general structures?

The domain of classical probability theory is the real line R^1. All the wellknown results like the central limit theorem, the law of the iterated logarithm or the law of large numbers concern real valued (or possibly vector valued in R^k) stochastic variables. The real line is so rich in structure that it can support the intricate but beautiful logical construction consisting of the probabilistic concepts and relations.

In mathematics in general there has been a trend towards generalization, abstraction, axiomatics. This is evident also in probability theory. Since Kolmogorov published his epoch-making *Grundbegriffe der Wahrscheinlich-keitsrechnung* we can use probabilistic arguments in completely general spaces without loosing anything in rigor. It is true that in such a generality we cannot always expect results of real mathematical substance, but the general framework is indispensable also for such, more concrete, work that is possible if the probability space is given more structure. The classical results indicate that such advance should be possible by defining *algebraic relations* in the space and studying their probabilistic implications. This leads us automatically to think of notions like groups, topological vector spaces and algebras. It is hard to resist posing the problem: does the classical probability theory have any counterpart in these more general algebraic structures? In the following chapters we shall see that sometimes this extension is made by an immediate and trivial generalization, sometimes a stronger effort is required leading to more profound discoveries, and sometimes we meet challenging problems to which the answers are known only partially if at all.

The fact that probability theory has grown so fast in recent years is due no doubt partly to its intrinsic value and its direct appeal to the mathematician but at least as much to its usefulness, exploited or potential. And so it is also with the present subject. Its motivation does not consist only of a wish to extend the theory to its natural boundaries. There are also a number of seemingly unrelated problems from physics, communication engineering, statistics and so on, that lead us to consider probabilistic relations in algebraic structures not equivalent to the real line (or the R^k-spaces). When more of these problems and results become widely known, we can expect an increased research activity and more rapid advance on both the practical and theoretical side of this subject.

We shall not jump directly into *medias res*. Instead let us start by recalling some fundamental facts and techniques from the classical theory (in 1.2.), then go on to review some problems posed by applications (in 1.3.), and finally (in 1.4.) sketch the outline of the historical development of the theory as it exists at present.

1.2. Classical methods and results

1.2.1. Granting the reader's permission we will take as our point of departure a brief sketch of some facts from elementary probability theory.

On the real line R^1 there are a number of basic probabilistic definitions that we will assume known as well as the corresponding fundamental relations,

    probability measure
    Borel measure
    stochastic variable
    independence
    various modes of convergence of stochastic variables
    (weak) convergence of probability distributions.

From the present point of view we are most interested in such relations that make use of the *additive* properties of R^1. Let P_1, P_2 be two probability measures, say defined through their distribution functions

$$F_i(y) = P_i\{x \,|\, x \leqslant y\}; \quad i = 1,2.$$

If to each P_i there corresponds a stochastic variable x_i; $i=1,2$; then the sum $x = x_1 + x_2$ has a probability distribution P given by a distribution function $F(x)$. If x_1 and x_2 are independent F is given as the *convolution* of F_1 and F_2

$$F(x) = \int_{-\infty}^{\infty} F_1(x-y)F_2(dy),$$

or shorter $F = F_1 * F_2$.—We also have the wellknown modifications of this formula if the distributions are absolutely continuous with respect to Lebesgue measure or with respect to counting measure on the set of integers; say the densities are

$$f(x) = \frac{F(dx)}{m(dx)} \text{ and } f_i(x) = \frac{F_i(dx)}{m(dx)},$$

then

$$f(x) = \int f_1(x-y) f_2(y) m(dy).$$

Note that m is in both of these cases a translation invariant measure.— The convolution operation is commutative, $F_1 * F_2 = F_2 * F_1$, and associative $(F_1 * F_2) * F_3 = F_1 * (F_2 * F_3)$. By iteration we can define $F_1 * F_2 * \ldots * F_n(x)$. The study of such convolutions, especially for large values of n, is one of the main tasks of probability theory.

There are some elementary relations. If the mean values exist

$$m_i = \int_{-\infty}^{\infty} x F_i(dx) = E x_i, \quad i = 1, 2, \ldots n,$$

then the mean value operation is additive

$$m = E x = m_1 + m_2 + \ldots + m_n.$$

If second moments exist then the variances

$$\text{Var}(x_i) = E(x_i - m_i)^2 = \sigma_i^2$$

satisfy

$$\text{Var}(x) = \sigma_1^2 + \sigma_2^2 + \ldots + \sigma_n^2,$$

if the stochastic variables x_i are independent. Under this hypothesis the variance is additive, and since a variance is non-negative the variance does not decrease when we add an independent stochastic element to a given one. In this sense (and in many others) convolution flattens out distributions. It also smoothes distributions, e.g. in the sense that if one of x_i has a continuous distribution then the same is true for the sum $x = x_1 + x_2 + \ldots + x_n$.

The most important analytical tool in this context is the *Fourier transform* or *characteristic function*

$$\hat{P}(z) = \varphi(z) = \int_{-\infty}^{\infty} e^{ixz} F(dx) = E \exp ixz, \quad z \text{ real}.$$

The importance of the characteristic function originates in its three properties

a) the characteristic function determines the probability measure uniquely

b) convolution corresponds to ordinary multiplication:

$$\text{if } P = P_1 \ast P_2 \text{ then } \hat{P} = \hat{P}_1 \cdot \hat{P}_2$$

c) weak convergence of a sequence of probability distributions to a limit distribution corresponds to convergence of the characteristic functions to a continuous limit function.

The *moments* of P are defined by

$$\alpha_k = Ex^k$$

if these integrals exist. Then they can be expressed as derivatives of the characteristic function

$$\alpha_k = 1/i^k \varphi^{(k)}(0).$$

Related concepts are the *semi-invariants* (or *cumulants*)

$$\gamma_k = 1/i^k \left(\frac{d^k}{dz^k} \log \varphi(z) \right)_{z=0}$$

which are of course linear combinations of moments of the same and lower order. The cumulants are additive for independent distributions. It is sometimes said that moments (and cumulants) are clumsy to work with and less generally defined than the characteristic function. This may be so in general investigations, but they are very useful in many cases, also when limit theorems are concerned.

1.2.2. Now let us go ahead to some limit theorems. One of the oldest is the *theorem of Bernoulli* going back to antiquity of probability theory: if x is a binomial variable $\nu = B(n,p)$ the relative frequency

$$p^* = \frac{\nu}{n}$$

converges in probability to the constant p. Or if we write ν as a sum of independent indicator variables $\nu = x_1 + x_2 + \ldots + x_n$, where $x_i = 1$ or 0 with probabilities p and $q = 1 - p$ respectively, then

$$p^* = \frac{1}{n}(x_1 + x_2 + \ldots + x_n) \to p = Ex_i.$$

Now we know of course that the above formula holds much more generally. This can be phrased in many different ways but the most attractive version of this *law of large numbers* is perhaps that due to Khinchin: if the x_i's are independent and identically distributed with a finite mean value m then

$$\bar{x} = \frac{1}{n}(x_1 + \ldots + x_n) \to m.$$

These asymptotic statements have been much sharpened. It is known (*de Moivre-Laplace*) that the distribution of the normed variable

$$\frac{v - np}{\sqrt{npq}}$$

converges to the normed, *normal (Gaussian)* distribution $N(0,1)$. Or, more generally, if the x_i's are independent and identically distributed with mean value m and finite variance $\sigma^2 > 0$ then the variables

$$\frac{x_1 + x_2 + \ldots + x_n - nm}{\sigma \sqrt{n}}.$$

again have $N(0,1)$ as their limit distribution. This is a simple form of the *central limit theorem*. Still more generally, if the x_i's are independent and with the same distribution (but with no restrictions upon its moments) what limit distribution can we get by appropriate norming of the sum x

$$\frac{x_1 + x_2 + \ldots + x_n - a_n}{b_n}?$$

It is known (Khinchin) that the possible limit distributions are the *stable ones*. These are the distributions such that if F is such a distribution function then to every $a_1 > 0$ and $a_2 > 0$ and arbitrary b_1 and b_2 there are constants $a > 0$ and b such that

$$F(a_1 x + b_1) * F(a_2 x + b_2) = F(ax + b).$$

The stable distributions are known in terms of their characteristic functions (see Notes 1.2.2.$_1$).

There are other limit theorems that do not fit directly into this scheme. Let us mention the familiar *Poisson* case, where the distribution of the binomial variable v tends to the Poisson distribution when $n \to \infty$, $np \to \lambda$. Here we do not keep the distribution of the x_i's fixed when n tends to infinity but vary it through the parameter p which is made smaller and smaller. To phrase this in more general terms we have to consider a *triangular array* of stochastic variables

$$\begin{cases} x_{11} \\ x_{21}, \; x_{22} \\ x_{31}, \; x_{32}, \; x_{33} \\ \quad \cdot \quad \cdot \quad \cdot \quad \cdot \quad \cdot \end{cases}$$

In each row we assume the variables to be independent and identically distributed. The second condition could be replaced by a condition that no term in the sum $x_{n1} + x_{n2} + \dots + x_{nn}$ makes a contribution of a size comparable with the whole sum. Otherwise we could of course get any limit distribution. Such a condition might be for instance that

$$\lim_{n \to \infty} \; \sup_{1 \leqslant k \leqslant n} P\{|x_{nk}| \geqslant \varepsilon\} = 0$$

for any $\varepsilon > 0$. Then the class of possible limit distributions coincides with the class of *infinitely divisible* distributions. A probability distribution P is said to be infinitely divisible if for any positive integer n there is a distribution P_n such that

$$P = P_n^{n*} = \underbrace{P_n * P_n * \dots * P_n}_{n \text{ times}}$$

A probability distribution is infinitely divisible if and only if its characteristic function is of the form (*the Lévy-Khinchin representation*)

$$\varphi(z) = \exp\left\{i\,az + \int \left[e^{iuz} - 1 - \frac{iuz}{1+u^2}\right] \frac{1+u^2}{u^2}\, G(du)\right\},$$

where a is a real constant and G is a non-decreasing function of bounded variation.

Criteria are known which enable us to obtain directly the asymptotic behaviour of sums of stochastic variables in terms of simple properties of the individual distribution functions. From a certain point of view this forms a complete theory which can be found expounded with admirable lucidity and precision in Kolmogorov-Gnedenko *"Limit distributions for sums of independent random variables"*—a *monumentum aere perennius* in the probabilistic literature.

1.2.3. It has turned out to be very fruitful to make a certain stochastic process correspond to a given limit theorem. This idea, that was put forward in a systematic manner already in Khinchin's *"Asymptotische Gesetze der Wahrscheinlichkeitsrechnung"*, is based on the concept of a *stochastic process of independent increments*. By this is meant a process $x(t)$ such that the increments $\xi_1 = x(t_2) - x(t_1)$, $\xi_2 = x(t_3) - x(t_2), \ldots \xi_n = x(t) - x(t_n)$ form independent stochastic variables for any t's such that $0 = t_1 < t_2 < \ldots < t_n < t$. Of particular importance are the *homogeneous* ones when the distribution P_h of any increment $x(t+h) - x(t)$, $h \geqslant 0$, depends only upon h (but not upon t). Such a process is said to be *continuous in probability* if $P_h \rightarrow \delta_0$ (weakly) as $h \downarrow 0$; by δ_x we always mean the degenerate distribution with all its mass in the point x.

Norming $x(0) = 0$ we can write $x(t)$ as the sum of independent stochastic variables

$$x(t) = \xi_1 + \xi_2 + \ldots + \xi_n.$$

If $x(t)$ has the properties defined above it is clear that $x(t)$ must have an infinitely divisible distribution for any $t > 0$. It is also easy to show that the characteristic function $\varphi(z,t)$ of $x(t)$ must be of the form

$$\varphi(z, t) = [\psi(z)]^t,$$

where $\psi(z)$ is a characteristic function (say expressed via the Lévy-Khinchin representation) of an infinitely divisible distribution, viz. that of $x(1)$.

Now the correspondence between the process and an appropriate limit theorem is intuitive although not so easy to establish with full rigor in every detail. The local behaviour of the sample functions of the process (continuity, size of the possible jumps etc.) will be governed by the function G in the Lévy-Khinchin representation. To mention just one possibility, if the sample functions are almost certainly continuous (assuming a separable version of the process) then G is constant except at $u = 0$ so that the characteristic function has the form

$$\varphi(z, t) = \exp t \left\{ imz - \frac{\sigma^2}{2} z^2 \right\}$$

which belongs to the normal process. The reader is referred to Doob [1] for a detailed discussion. The Poisson process is another easily treated example.

On the other hand the function G of a limit law is known to be expressible through the distribution functions of the summands x_{nk}. This establishes a logical connection between the behavior of the asymptotically infinitesimal summands of the limit theorem and the local properties of the sample functions of the associated stochastic process of independent increments.

Among the mathematical techniques that have been used to prove the classical limit theorems Fourier analysis is of course the most successful in general, but in some special instances it is simpler to work with the related Laplace transform, moments etc. A different approach is that used to establish the correspondence discussed in this section. One derives the functional equation governing the behavior of the distributions of the associated process. This may be the heat equation, the Fokker-Plank equation or something more general. Then one considers the distributions of the cumulative sums in question, tries to show that they satisfy a functional equation that is close in some sense to the first equation and finally a continuity argument is applied (see Notes 1.2.2.$_2$).

In the following chapters much attention is given to the development of techniques appropriate for the study of limit theorems for independent variables and of processes of independent increments. We shall then be guided all the time by what we know from the classical theory.

1.3. Practical background to the theory

Although not immune to the aesthetic appeal of a general and well rounded off mathematical theory the author believes that the strongest incitement to continued work in the probability theory of algebraic structures will come from the applications. Already now the variety and number of practical situations leading to problems of the sort we have in mind indicate the need of such a theory. Here we can only sketch a few typical cases.

1.3.1. Say that we have at our disposal a method of generating random numbers. The numbers obtained are given in the form

$$x = 0.x_1 x_2 ... x_k$$

in decimal notation. To each of the 10^k different possible numbers belongs a certain probability, defining a probability distribution that may be only partially known to us. We now let the device generate a sequence of independent numbers $x^{(1)}$, $x^{(2)}$,...,$x^{(n)}$ and combine these in the following way. We add the numbers and retain only its fractional part ξ_n,

$$x^{(1)} + x^{(2)} + ... + x^{(n)} = \text{integer} + \xi_n.$$

If n is large we may hope that the distribution of ξ_n stabilizes around some uniform distribution (over the possible values) which can be considered as known to us.

Here we obviously have to deal with the *cyclic group Z* of order 10^k.

This problem is so simple that one can solve it directly by purely elementary methods. However in such a solution would be hidden the germ of a more general idea that would be worth pursuing further.

1.3.2. Let ξ_t be a stationary normal sequence with mean values zero and covariances $r_h = E\xi_t\xi_{t+h}$; $t, h = 0, \pm 1, \pm 2, \ldots$ If ξ_t is passed through a filter acting instantaneously on the input, so that the output is just some well-behaved function $g(\xi_t)$ of ξ_t, then we can study the properties of the output. As is often the case for stationary processes we may prefer to work with the spectra rather than with the covariances.

Computations give us the spectral distribution function of the output as a certain linear combination of the spectral distribution functions associated with the covariances $r_h^\nu, \nu = 0, 1, 2, \ldots$ But in frequency language this amounts to forming iterated convolutions of the original spectral distribution function $F(\lambda)$

$$\begin{cases} r_h \leftrightarrow F(\lambda) \\ r_h^2 \leftrightarrow F*F(\lambda) \\ r_h^3 \leftrightarrow F*F*F(\lambda) \\ \quad \cdot \quad \cdot \quad \cdot \quad \cdot \quad \cdot \quad \cdot \end{cases}$$

While we may be able and willing to compute a few of these convolutions directly it is clear that we could use an approximation for the higher order ones. We can write $F = \sigma^2 \cdot G$, where G is an ordinary (normed) distribution function and σ^2 is the variance of the input. Then we have

$$F^{n*} = \sigma^{2n} \cdot G^{n*}.$$

But in the convolution integrals defining F^{n*} and G^{n*} we identify frequencies differing by a multiple of 2π. In other words we deal with addition on the torus group T^1, G defines a probability distribution on this group and G^{n*} is the result after "adding" n independently and identically distributed group elements. We will get a useful approximation if we can show that G^{n*} converges to some limit distribution. Again it is almost obvious that such a limit distribution, if it exists, must have some uniform (i.e. invariance) property (see Notes 1.3.2).

The above can be repeated almost word for word on the higher dimensional torus groups T^k; in principle the solution is just as simple for larger values of k.

Note that so far we have only met commutative groups.

1.3.3. Consider now a physical system (liquid, gas …) with its particles performing a Brownian motion. For concreteness let us assume that there is no mean value drift and that the motion is isotropic. In the system is immersed a sphere with its center fixed whose motion is coupled through friction forces to that of the surrounding medium (see Notes 1.3.3).

In a given time interval (t_1, t_2) the sphere rotates and we shall call the rotation 0_1. It may be instructive to think of one fixed coordinate system and another mobile one moving together with the sphere. In coordinates 0_1 could be expressed as an orthogonal matrix. Anyway 0_1 is a random rotation. During the next time interval (t_2, t_3) we get a new rotation 0_2, stochastically independent of the first under certain conditions (lack of inertia ...), and so on $0_3, 0_4,...$ and finally 0_n under the time interval (t_n, t_{n+1}). The total rotation during the time (t_1, t_{n+1}) is then $R_n = 0_1 0_2 ... 0_n$. To study the behaviour of R_n for large values of n we are led to consider probabilistic limit laws on the *orthogonal group*. Here we meet with something interesting. In the previous examples the groups were commutative but in the composition of rotations the order of performing the operations is essential. This *non-commutative* property is responsible for a corresponding complication in the mathematical technique that will be used.

Here the Brownian motion acted upon a very simple physical concept, viz. a sphere. It is not difficult to think of realistic situations where the simple sphere is replaced by another solid, an elastic body or where we have electrical systems instead of mechanical ones. As long as we can write down differential equations governing the systems we will run into similar although perhaps more complicated problems. Our object will be a *Lie group*, our objective to introduce a probability structure on such groups. As long as these groups are *compact* we can expect to find limiting (uniform) distributions and the reader may start thinking of *Haar measure*. Then he is right; this concept will play a fundamental role here. Of course it is not really essential for that statement that the underlying groups are Lie groups, the important thing is that they be compact.

1.3.4. A *disordered linear lattice* is a one dimensional chain of particles (which word can be given a wide interpretation) where contiguous particles are connected by elastic forces and where the particles or forces have random characteristica. Often it is possible to think of one particle acting upon the next one in the chain through a *transfer matrix*. In the simplest case these will be 2×2 matrices. They will be combined through ordinary matrix multiplication. If these *stochastic matrices* are independent of each other then the set-up looks very much like the one just discussed. However, if one carries out the computations it will be seen that the groups involved will no longer always be compact. This leads to a drastic increase in the degree of mathematical sophistication when we try to find mathematical tools useful to solve the stochastic problem. As we shall see later there are new and fundamental problems already in *formulating* the possible limit theorems (see Notes 1.3.4).

Let us think of a chain of devices that can be of fairly general physical nature. If the effect of such a device can be expressed through a transfer matrix we still have something similar. Say that the chain is governed by a linear difference equation

$$u_{n+p} + a_n^{(p-1)} u_{n+p-1} + \ldots + a_n^{(0)} u_n = v_n,$$

where the right member can be deterministic or random, possibly identically zero, and where the coefficients $a_n^{(j)}$ are random. Introduce the vectors

$$\xi_n = \left\{ \begin{array}{c} u_{n+p} \\ u_{n+p-1} \\ \vdots \\ u_{n+1} \\ u_n \end{array} \right\}$$

Then $\xi_{n+1} = M_n \xi_n + \eta_n$, where M_n is a stochastic matrix. Especially if the η's vanish we get a model related directly to the previous one.

Since the stochastic properties of the transfer matrix can be chosen in many different ways one can expect a great variety of different although related results. As an example let us mention the case of wave guides with stochastic inhomogeneities. Approximating it as a discrete chain it is possible to describe it in terms of the (complex) reflection coefficients. It turns out that we are led to study probability distributions in the open unit circle $|z| < 1$ and where the law of composition $\circ$ is given by

$$z_1 \circ z_2 = \frac{z_1 + z_2}{1 + \bar{z}_1 z_2}.$$

But the set of linear fractional transformations

$$\frac{z_1 + z}{1 + \bar{z}_1 z}$$

is the well-known group of transformations leaving the unit circle invariant. Our problem will hence be of a form similar to the earlier ones, to study the probability distribution of the composition $z_1 \circ z_2 \circ \ldots \circ z_n$ for a large number of independent and identically distributed group elements z_ν.

If the difference equation is not linear the matrices must be replaced by non-linear operators. We may still be led to stochastic groups at least if these operators are all non-singular. But it may also be relevant to consider them as elements of a *stochastic algebra*, and this will have some very interesting implications as we shall see in a later chapter. It is too early to discuss these yet but it will be sufficient to point out the intriguing problem of determining the spectral properties of the elements of such stochastic algebras.

Depending upon the time- and space-scale of the physical problem a differential equation may provide a better description of the physical set-up than the difference equation above. The probabilistic counterpart will then be stochastic processes taking values in some group, algebra etc. and probably of "independent increments". The infinitesimal transformations will be of relevance especially since they will have simple and concrete physical interpretations.

Note that so far all the domains of probability distributions have been *locally compact*.

1.3.5. Let us now turn to a problem in signal detection. We observe a stochastic process $x(t)$, $t \in (0, T)$. To simplify things we assume that $x(t)$ forms a normal process with known covariance function but with two hypothetical mean value functions $H_0 : Ex(t) \equiv 0$, $H_1 : Ex(t) = m(t)$; $t \in (0, T)$. The null hypothesis corresponds to no signal present, only noise, while under the alternative hypothesis $x(t)$ contains a signal component. We wish to find a criterion to distinguish the two hypotheses from each other.

This sort of statistical inference is nowadays dealt with as follows. The two hypotheses correspond to two probability measures P_0 and P_1 on some probability space Ω. Compute the Radon-Nikodym derivative $p(\omega) = P_1(d\omega)/P_0(d\omega)$ if P_1 is absolutely continuous with respect to P_1 and form the critical region à la Neyman-Pearson

$$W = \{\omega \,|\, p(\omega) \geqslant c\},$$

where the constant c is chosen so that the probability $P_0(W)$ has a reasonable value. In the present case we have a linear problem at hand and it is possible to find $p(\omega)$ by a simple application of Hilbert space methods. In such an approach we consider the process as a continuum of stochastic variables, each represented as a point in a Hilbert space. Hence we do not really specify the individual sample functions. From a certain point of view it would be more attractive (although perhaps not more practical) to start from the individual sample functions, operate upon them by a suitable device and so obtain a criterion for signal detection.

Now we can do this by assuming that the sample functions together form some functional space. Perhaps we use some *Banach space*. Then we have to introduce probability measures on such spaces and study the properties of such measures. Note that these spaces are in general not locally compact (which makes for certain complications), but on the other hand they have instead a linear structure (which certainly simplifies things).

To show more clearly what sort of things can be done via this approach let us assume that we sample repeatedly from the same process as before. Denote the observed sample functions $x_1(t), x_2(t), \ldots x_n(t)$ and form the average

$$\bar{x}_n(t) = \frac{1}{n}\left[x_1(t) + x_2(t) + \ldots + x_n(t)\right].$$

This average also belongs to the linear space and we can ask if it converges (in the topology of the same space) when n increases. Of course we would like it to converge to the mean value function $m(t) = Ex(t)$, and we would like to prove (strong) convergence in probability

$$\lim_{n\to\infty} P\left\{\|m(t) - \bar{x}_n(t)\| > \varepsilon\right\} = 0, \quad \text{arbitrary } \varepsilon > 0,$$

or (strong) convergence almost certainly

$$P\left\{\lim_{n\to\infty}\|m(t) - \bar{x}_n(t)\| = 0\right\} = 1,$$

or perhaps some other type of convergence. Such *laws of large numbers in linear spaces* would certainly be useful.

In the same way one would like to have central limit theorems etc. in stochastic linear spaces. One example will be enough. Say that we sample from a continuous population on the real line with known distribution function. It is known that we can transform this to a rectangular distribution on the unit interval $(0, 1)$. Form the empirical distribution function

$$F_n(x) = \frac{1}{n}\,[\text{number of observations } \leqslant x].$$

Then the function

$$y_n(x) = \sqrt{n}[F_n(x) - x]$$

can be considered as an element of the Hilbert space $L_2(0, 1)$ spanned by all functions quadratically integrable over $(0, 1)$. This defines a probability distribution P_n on this space $L_2(0,1)$. When n increases we would like to prove that P_n converges (in what sense?) to a limiting distribution P. It would then be of great interest to be able to infer that

$$P_n\{|y_n(x)| \leqslant c\} \to P\{|y(x)| \leqslant c\};$$

this is of course related to the tests of Kolmogorov and Smirnov (see also Notes 1.2.2.$_2$).

1.3.6. At the end of section 1.3.4. we discussed stochastic algebras. Then we had only locally compact algebras in mind but now it is natural to think of *Banach algebras*, combining the linear structure of the probability space with that of an algebra. This will be the natural thing to do in such applications where a bounded stochastic operator acts linearly on a Banach space. Unbounded operators, that will appear e.g. in connection with stochastic differential operators, fall outside of our framework.

1.4. Historical background

1.4.1. It is always difficult to trace mathematical ideas back to their beginning; with some effort and goodwill this search for something vague and ill defined can often be carried out with success very far back in time. The present subject is no exception to this rule. Although most of the theory of to-day dates from the 1950's and especially from the last few years some of the ideas originate much earlier.

In the classical probability theory of limit theorems etc. the domain was the real line or a lattice embedded in it but it became apparent in the early development of the theory that it could often be extended to the plane (R^2) or higher dimensional Euclidean spaces (R^k). Take for example the law of large numbers. Once it has been established for real valued stochastic variables we can state and prove in practically the same way a vector valued version of it. Very little new mathematical substance is added while we do this. Other limit theorems may be a bit more difficult to generalize to R^k but in general the same is true: it is not very difficult, perhaps not very interesting but sometimes quite useful.

1.4.2. The situation becomes quite different when we come to infinite dimensional vector spaces. Here a more serious effort is needed to carry over the classical results or to find the new results taking the place of the already known ones. In a series of papers Fréchet ([1] and others) has argued that one should study *probability theory in topological spaces* of varying generality. His suggestion—very natural to come from one of the creators of modern functional analysis—did not meet with much response to begin with and it has taken some time before probabilists in general became aware of the possibilities in this direction.

An exception to this slow development can be found in an early paper by Kolmogorov [2]. Already in 1935 he proposed to study probability theory in Banach spaces by the aid of what he called the *characteristic functional*. Let X be the Banach space with elements x, X^* the dual space of linear (bounded) functionals $x^* = x^*(x)$ and P a probability distribution over X. For the moment we leave aside the question of how P should be introduced in the space. Kolmogorov defines the characteristic functional as

$$\hat{P}(x^*) = E \exp i x^*(x);$$

it is a complex valued function on X^*. It has at least some of the main properties of the characteristic function. It defines the probability distribution uniquely (uniqueness theorem), convolution corresponds to ordinary multiplication of complex valued functions and it is positive definite. But there are other questions, such as the analogue of the continuity theorem, that lie considerably deeper and remained open for some time.

The notion of characteristic functional has been suggested in different contexts by a number of authors, Bochner [1], Le Cam [1] and others. It is now an indispensable tool in stochastic linear spaces.

1.4.3. The systematic attack on these problems was begun around 1953. In a fundamental paper of Mourier [1] Fourier analysis was studied in more detail and a great stride forward was taken in the theory of limit theorems. Several versions of the law of large numbers were derived and a partial result was obtained for the central limit theorem in a Hilbert space. The latter was completed in a paper of Mourier and Fortet [2]. The normal distribution was defined in different but equivalent ways, e.g. as a distribution such that all the linear functionals have normal (real or complex) distributions.

Mourier also considered the measure-theoretic set up that should be used in a Banach space. Probability distributions were introduced via all the (scalar) stochastic variables $x^*(x)$ when x^* runs through the dual X^*. To make the functions ordinary stochastic variables they were assumed to be measurable. A slightly different approach was used by Hanš [1], who assumed instead that P was a Borel measure. While Mourier based her probability distributions on the half-spaces $\{x \mid x^*(x) \leqslant c\} \subset X$, Hanš used the open sets as building blocks. For a separable space these two methods give the same result and it is only in the non-separable space that the difference is essential.

Once the linear functionals are measurable one can define the *expected value* of a stochastic element in a Banach space as follows (see Notes 1.4.3). If there is an element $m \in X$ such that the equation

$$x^*(m) = \int_X x^*(x)P(dx)$$

has a sense and is valid for every $x^* \in X^*$, then m is called the expected value of P and denoted $m = Ex$. This is nothing but the Pettis integral applied to the present context. Many of the properties of the ordinary expected value hold in a Banach space. This definition is necessary for the law of large numbers.

Variances, covariances (or rather covariance operators) can be defined in a stochastic Hilbert space. Higher moments do not seem to have been used.

Now the foundation had been laid of a theory of *stochastic Banach spaces* but it had at least two irritating gaps. One was the already mentioned difficulty that appeared when one tried to extend the continuity theorem. Say we have a sequence of probability measures $P_1, P_2, \ldots$ on X. In the usual terminology P_n *converges weakly* to a limit distribution P if

$$\lim_{n \to \infty} \int_X f(x)P_n(dx) = \int_X f(x)P(dx)$$

for every continuous and bounded complex valued function $f(x)$. If this is so then of course $\hat{P}_n(x^*) \to \hat{P}(x^*)$, $x^* \in X^*$, where "$\wedge$" denotes the Fourier transform, the characteristic functional. This is seen just by putting $f(x) = \exp i x^*(x)$. Instead, if we assume that the $\hat{P}_n(x^*)$ converge, can we then infer that the P_n converge weakly to a probability distribution? Simple examples show that this is not so in general and the reason for this unpleasant fact is that probability mass can flow out to higher and higher dimensions (think of a Hilbert space with denumerably many coordinate axes). To counteract this some compactness assumption is needed on the P_n's. This can be done in many ways, and especially for Hilbert spaces we do have some adequate criteria due to Prohorov [1].

The other gap is the following one. Suppose that $\varphi(x^*)$ is a continuous, complex function on X^*, continuous in x^*, $\varphi(0) = 1$, $\varphi(-x^*) = \overline{\varphi(x^*)}$ and positive definite. Does there exist a probability distribution P on X such that $\varphi(x^*) = \hat{P}(x^*)$? Again the answer is: not always. Now one may suspect that the critical word in the statement of this problem is "continuous". We have interpreted it as "strongly continuous". If we instead use weak convergence we still do not get a true statement. Instead it was shown for Hilbert spaces by Kolmogorov [3] that another sort of convergence ("S-") should be used and this gives a partial answer to the problem. Another, more general but somewhat difficult, condition has been given by Mourier [1].

This theory is in rapid development.

1.4.4. Other spaces can be given probability structures in a meaningful way. If the space possesses some linear properties things are similar to the Banach space situation. Gelfand [1] and K. Îto [2] have introduced probability measures in the set of *Schwartz distributions* and this work has been continued by Ullrich [1] and others. Of course Fourier analysis will still be the main tool.

Another line of development has been in the direction of general metric spaces. While it is still possible to define probability distributions and even mean values etc. in such spaces it is clear that we will not get results like these discussed above unless addition or some other binary operation is defined. Therefore such work falls outside the scope of this book and we shall not discuss it further.

1.4.5. If any concept can be singled out as central in our subject it must be that of *stochastic group*. The linear spaces are of course groups under addition but also possess many other properties. It is possible to get other, valuable results by leaving out the linearity and instead assuming some (local) compactness property of the underlying topological group.

There are some isolated but important early studies of this sort in the literature. Already in 1928 Perrin [1] was led by physical considerations (see section 1.3.3) to consider a stochastic group on the group of rotations of 3-space. In spite of the special nature of this investigation it contained ideas of general scope. The same can be said about a paper of Lévy [1] containing a detailed discussion of the circle group.

The break-through came in 1940 in the paper *On the probability distribution on a compact group* by K. Îto and Kawada. This paper that seems to have gone almost unnoticed until recently is noteworthy both because of its results and especially because of its technique. These authors discussed what happens with convolutions P^{n*} on a compact group and they found that the possible limit distributions are formed by the Haar measures on certain subgroups. To prove this they used Fourier analysis leaning heavily on the theory of unitary representations of compact groups. Consider the whole set of irreducible, unitary representations

$$U_0(g) \equiv I, \ U_1(g), \ U_2(g), \ \ldots$$

All these $U_j(g)$ can be given as unitary matrices in finite dimensional spaces. The *Fourier transform* of P is then defined as

$$\hat{P}_j = \int_G U_j(g)P(dg); \quad j = 0, 1, 2, \ldots$$

Since this Fourier transform behaves very much like the characteristic function it is well suited to prove limit theorems and so on as Îto and Kawada did.

These findings have been rediscovered, modified or extended by a number of authors among whom one should mention especially Stromberg [1], Kloss [1] and Urbanik [1]. The finite groups have been treated by Vorobev [1], Dvoretzky-Wolfowitz [1] and by Böge [1]; the latter author investigates the form of the corresponding infinitely divisible laws.

To-day we have a theory of compact stochastic groups; perhaps it cannot yet be said to be complete, but it answers some of the main problems. In retrospect it may be thought that the compactness facilitates the problems so much that they should be solvable without access to any elaborate analytical technique. However this would be to underestimate the original conceptual difficulties. Also these results indicate the way for further progress.

1.4.6. If we assume that the group is only locally compact we meet additional difficulties. The concept of Fourier transform is generalized easily enough but its basic properties are not proved quite automatically. Results in this direction were given in Grenander [3], [4], as well as certain results obtained by Fourier analysis.

For this purpose he used the theory (due to Godement, Naimark and others) of representations of locally compact groups. This led to theorems similar to the law of large numbers and the central limit theorem but it was also shown that there are simple situations where the result does not imitate what happens in the classical theory. In this fairly general situation our knowledge is less complete but the outline of a theory can be distinguished.

The operator

$$Tf(g) = \int_G f(gh)P(dh)$$

plays an important role. Its spectral properties have been studied by Grenander [2] and Kesten [1].

One particularly interesting case is that of a Lie group. It was shown already in 1956 by Hunt that the homogeneous processes on such a group can be characterized completely. Hunt did this by deriving the general expression for the *infinitesimal generator*

$$A f(g) = \lim_{h \downarrow 0} \frac{T_h - I}{h} f(g),$$

where
$$T_t f(g) = \int_G f(gh) P_t(dh),$$

and where the function $f(g)$ is smooth enough. By similar methods Wehn [1] showed that a number of important limit theorems could be obtained.

Returning to the general case we would like to stress the importance of the concept of an *idempotent measure*. A probability measure M is said to be idempotent if $M^{2*} = M * M = M$. Let P be a probability measure on the group and assume that Q is a limit measure, $P^{n*} \to Q$. Then $Q^{2*} = \lim_n P^{n*} * P^{n*} = \lim_n P^{n*} = Q$; Q must be idempotent. Also any idempotent Q is a limit measure $Q = Q^{n*} = \lim_{n \to \infty} Q^{n*}$. The class of possible limit distributions coincides with the class of idempotent distributions which makes it clear why these are so important to us. The idempotents have been studied under various assumptions about the underlying algebraic topological structure by Wendel [1] and others. Typically an idempotent is a normed Haar measure on some compact subgroup.

If Q is idempotent then its support, i.e. the smallest set containing all the mass (a more careful definition is given later on), must be a subgroup. To make this plausible think of a finite distribution with point masses p_g, $g \in G$. We have

$$p_g = \sum_{ab=g} p_a p_b.$$

Introduce the set $\Gamma = \{g \,|\, p_g > 0\}$. If p_a, $p_b > 0$ then $p_{ab} > 0$ so that $\Gamma^2 \subset \Gamma$. But since Γ is finite this implies that Γ forms a subgroup. We shall see in the main text that this holds more generally.

A type of groups with more in common with the linear spaces are those with *unique n^{th} roots*: to every integer n and every element $g \in G$ there should correspond a uniquely determined element $h \in G$ such that $h^n = g$. Such stochastic groups were considered by Grenander [4]. On such groups the limit theorems take on a particularly attractive form.

1.4.7. Some authors have studied *stochastic semi-groups* and their investigations indicate where and how much the whole group property is essential for the validity of the results discussed above. So far the compact semi-groups are the only ones treated in some detail.

Hewitt-Zuckerman [1] dealt with finite commutative semi-groups by using semi-characters for their definition of the Fourier transform of a probability distribution. A semicharacter $x(s)$ is a complex valued function $x(s) \not\equiv 0$ such that $x(st) = x(s)x(t)$ for all elements s and t of the semigroup. The Fourier transform is then defined as

$$\hat{P}(x) = \sum_s p_s x(s)$$

and it has some of the familiar properties of Fourier transforms. These authors gave among other things a criterion for P^{n*} to converge.

Rosenblatt [1], [2] investigated limit theorems on compact semi-groups and showed among other things that the averages

$$\pi_n = \frac{1}{n}(P + P^{2*} + \ldots + P^{n*})$$

always converge to some limiting measure π as n tends to infinity. π is idempotent and has as its support the minimum two-sided ideal of the semi-group.

It is easy to give a heuristic motivation why the kernel ($=$minimum two-sided ideal) enters naturally when we study limit theorems and idempotent measures just as the subgroups appear in the group case. Suppose that P^{n*} converges to Q. The stochastic element $\gamma_n = g_1 g_2 \ldots g_n$ will take values in the support $s(P^{n*})$ and we can confine ourselves to the closed semi-group S spanned by these supports, $n = 1, 2, \ldots$ If I is a two-sided ideal of S and n is large then γ_n will take as values elements in S close to I. But once we are in I we will stay there. Now consider γ_{rn}

$$\gamma_{rn} = g_1 \ldots g_n g_{n+1} \ldots g_{2n} \ldots g_{rn-1} \ldots g_{rn}$$

split into r long blocks. With overwhelming probability (when r is also large) at least one of these blocks represents an element close to I. Therefore we can expect the limit distribution to have its support in I; we can perhaps hope that $s(Q)$ will be the smallest I, i.e. K.

1.4.8. Another direction of research has been to the stochastic algebras. We then have the two main operations, addition and multiplication. Grenander [2], [3] investigated the relation between additive and multiplicative processes and the consequences of this for the relevant limit theorems. If, more particularly, we have a Banach algebra as our object, then we get a natural connection with what we know about probability theory on Banach spaces.

An important special case is that of $k \times k$ matrices. Kesten-Furstenberg [1] have shown that the limit

$$\lim_{n\to\infty} \frac{1}{n} E \log \|M_1 M_2 \ldots M_n\|$$

exists, and that if it is finite then it coincides with the limit

$$\lim_{n\to\infty} \frac{1}{n} \log \|M_1 M_2 \ldots M_n\|$$

existing almost certainly. The norm is here defined as usual

$$\|M\| = \max_i \sum_j |m_{ij}|, \quad M = \{m_{ij}\}.$$

Under a more restrictive hypothesis they also derive the asymptotic distribution of the elements $\log m_{ij}^{(n)}, N_n = \{m_{ij}^{(n)}\} = M_1 M_2 \ldots M_n$. In spite of the high specialization of these results they merit careful attention as suggestive for further work.

In this connection one should mention probabilistic functional analysis in general, say the study of stochastic functional equations, their iterative solutions and so on. Such problems have been studied by Špaček, Hanš, Bharucha-Reid and several other authors.

We have now got a birds-eye-view of the theory and what motivates it. We are ready to go to work.

CHAPTER 2

STOCHASTIC SEMI-GROUPS

2.1. Generalities

The probability measures to be considered in this book will have a separable topological Hausdorff space, say S, as their domain (see Notes 2.1 for the definition of separability). A probability measure P should be defined on the sets belonging to the σ-algebra $\mathcal{A}$ generated by all open sets in S (see Notes 2.1.$_1$). Occasionally we shall also use measures m that are not bounded, $m(S)$ not finite; some of the statements below will be phrased with this in mind.

We shall use some of the terminology familiar to us from the probability theory on finite dimensional Euclidean spaces, and some of the notions are carried over to the present situation immediately. Two stochastic elements in S are said to be independent if their joint distribution over $S \times S$ is a product measure; similarly for several stochastic elements. A probability distribution P is said to be continuous if $P\{s\}=0$ for every set $\{s\}$ consisting of a single element s. It is said to be discrete if all its mass is concentrated on elements $s_1, s_2, \ldots$ with point masses $p_1, p_2, \ldots p_n, \ldots \geqslant 0$, $\sum_1^\infty p_n = 1$. For a given σ-finite measure m we say that P is absolutely continuous with respect to m if

$$P(E) = \int_E p(s)m(ds) \ \text{ for all } \ E \in \mathcal{A};$$

$p(s)$ is called the frequency function of P.

Much attention will be given to locally compact spaces and we consider the Borel measures defined on the σ-algebra generated by all compact sets. It is then convenient and not very restrictive to assume that the probability measure is regular, i.e.

$$\begin{cases} P(E) = \inf P(O), & O = \text{open set covering } E \\ P(E) = \sup P(C), & C = \text{compact set contained in } E. \end{cases}$$

In the present context the regularity assumption is automatically fulfilled, as pointed out in Notes 2.1.$_1$. It should be noted that in the non-separable case this is no longer so.

If our probability measure has no mass outside a certain part of S we may wish to study only what happens in such a part, chosen as small as possible. This is done through the notion of the *support*, $s(P)$, of the measure P. Among the possible alternative definitions of the support we choose the following: $s(P)$ is defined as the set of those points s such that every neighbourhood of s has positive measure. The support is a closed subset of the whole space S.

For a measure m of the type (regular, Borel) described we can form the integral

$$I(f) = \int_S f(s)m(ds),$$

at least for certain real valued functions $f(s)$. It is certainly defined for $f \in L(S) =$ the set of all continuous real functions on S, vanishing outside of compact sets. $I(f)$ is a linear functional on $L(S)$; furthermore it is a positive functional: $I(f) \geqslant 0$ if $f \in L^+(S) =$ the set of functions belonging to $L(S)$ and which are non-negative. This statement can be inverted. If a given functional $I(f)$ defined on $L(S)$ has these properties then there exists a uniquely determined (regular), Borel measure m such that the above integral representation holds (see Notes 2.1.$_2$).

Now introduce the norm

$$\|f\| = \sup_{s \in S} |f(s)|$$

among all continuous and complex valued functions $f(s)$ vanishing at infinity. The bounded linear functionals on the resulting Banach space $L_\omega(S)$ can then be represented as

$$\mu(f) = \int_S f(s)\,\mu(ds),$$

where μ is a (regular) complex measure, and the norm $\|\mu\|$ of the functional is given by the total (absolute) variation of the measure μ. (Notes 2.1.$_2$.)

During our search for limit theorems it will be necessary to specify the type of convergence of probability distributions. The two main types are vague and weak convergence; it should be noted that the terminology varies a good deal in the literature. A sequence of probability measures $P_1, P_2, P_3, \ldots$ on S could be said to *converge vaguely* towards the measure m if

$$\int_S f(s)P_n(ds) \to \int_S f(s)m(ds) \tag{1}$$

as n tends to infinity for every $f \in L(S)$. However we shall more often use the following definition. If (1) holds for every bounded and continuous f we speak of *weak convergence*. The latter obviously implies that the limit, m, is a probability measure. The two concepts coincide if S is compact. In general a sequence of probability measures P_n converges weakly to a limiting measure Q if and only if it converges vaguely and if $Q(S) = 1$. The "only if" part of the statement is obvious. To prove the converse we note that if $P_n \to Q$ vaguely and if $Q(S) = 1$, then for any $\varepsilon > 0$ we can find a compact set C_ε with $Q(C_\varepsilon)$ and $P_n(C_\varepsilon) > 1 - \varepsilon$ for all n. But if $f(s)$ is an arbitrary bounded and continuous function we can approximate it uniformly on C_ε by some $f_\varepsilon \in L(S)$. Hence

$$\int_S f(s) P_n(ds) - \int_S f(s) Q(ds) = \int_S f_\varepsilon(s)[P_n(ds) - Q(ds)]$$

$$+ \int_{C_\varepsilon} [f(s) - f_\varepsilon(s)][P_n(ds) - Q(ds)] + \int_{C_\varepsilon^*} [f(s) - f_\varepsilon(s)][P_n(ds) - Q(ds)]$$

where each of the three terms on the right can be made arbitrarily small. Let us also note that $\mathcal{P}(S)$, the set of regular, Borel probability measures over S, is compact in the vague topology, see Notes 2.1.$_3$.

We shall use throughout the notation δ_s for the degenerate probability measure with all the mass at the element $s \in S$. We have in particular convergence in probability to a fixed element, $P_n \to \delta_s$, $s \in S$.

2.2. Stochastic semi-groups

Now we will let the space S also have an algebraic structure and assume that it forms a *topological semi-group*. On S there is defined a binary operation, multiplication, which should be a continuous function of both arguments. It should also be associative but not necessarily commutative. As in the last section we can introduce (regular), Borel probability measures in S and our first task is to put the probabilistic and algebraic properties of S into relation with each other.

Say that we observe two stochastic elements s and t from the populations P and Q respectively on the locally compact semi-group S. Form the new stochastic element $u = st$; we want to be able to make statements about its probability distribution. To define its distribution consider the integral

$$R(f) = \int_S \int_S f(st) P(ds) Q(dt) \tag{1}$$

where the notation should mean that the integral is taken with respect to the product measure $P \times Q$. This functional is defined for every $f \in L(S)$ and is positive.

Definition 2.2. By the convolution $P \ast Q = R$ we mean the uniquely determined (regular), probability measure R satisfying

$$R(f) = \int_S f(s)\, R(ds).$$

It is possible to express the convolution $P \ast Q$ a little more directly.

Theorem 2.2.1. The convolution can be written as

$$R(B) = P \ast Q(B) = \int_S \int_S I_B(st) P(ds) Q(dt) = T(C) \tag{2}$$

for any Borel set B of S, and where $I_B(s)$ is the indicator function of B.

Proof: It is enough to show $R(C) = T(C)$ for any compact set $C \subset S$, since the regular measures are determined by their values on compact sets. Let us cover C by an open set 0, $C \subset 0$, such that $R(0) < R(C) + \varepsilon$, $T(0) < T(C) + \varepsilon$ for a given positive ε. Introduce the function

$$f(s) = \begin{cases} 1 \text{ if } s \in C \\ \text{continuous for all } s \\ 0 \text{ if } s \notin 0 \end{cases}$$

Then
$$R(C) \leqslant \int_S f(s) R(ds) = \int_S \int_S f(st) P(ds)\, Q(dt)$$

$$\leqslant \int_S \int_S I_0(st) P(ds) Q(dt) = T(0) < T(C) + \varepsilon.$$

Similarly
$$T(C) = \int_S \int_S I_C(st) P(ds) Q(dt) \leqslant \int_S \int_S f(st) P(ds) Q(dt)$$

$$= \int_S f(s) R(ds) \leqslant R(0) < R(C) + \varepsilon,$$

and the two inequalities obtained prove the statement. (Notes 2.2.$_{.1}$.)

Remark. If S is also a topological *group* we can simplify a bit more since obviously

$$P \ast Q(B) = T(B) = \int_S P(Bt^{-1})Q(dt) = \int_S Q(s^{-1}B)P(ds)$$

We shall often use the notation $\mathcal{P}(S)$ for the set of all Borel probability measures on S. It is not too difficult to show that $\mathcal{P}(S)$ is a topological semi-group with convolution as the binary operation (see Notes 2.2.$_3$). Also one can show that $\mathcal{M}(S) =$ the set of all bounded and Borel, complex measures forms a Banach algebra with the usual definition of addition and scalar multiplication, multiplication defined as convolution and the norm taken as the total absolute variation.

We have a useful and simple expression for the support of the convolution $s(P \ast Q)$.

Theorem 2.2.2. The support of the convolution $P \ast Q$ is the closure of the product of the supports of P and Q respectively, $s(P \ast Q) = \overline{s(P) \cdot s(Q)}$.

Proof: To prove that $s(P \ast Q) \subset \overline{s(P) \cdot s(Q)} = A$ note that

$$P \ast Q(A) \geqslant \int_S \int_S I_{s(P)s(Q)}(st)P(ds)Q(dt) \geqslant \int_S I_{s(P)}(s)P(ds) \int_S I_{s(Q)}(t)Q(dt) = 1$$

so that $P \ast Q\,[\overline{s(P)s(Q)}] = 1$ and $s(P \ast Q) \subset \overline{s(P)s(Q)}$.

To prove that $s(P \ast Q) \supset \overline{s(P)s(Q)}$ let us choose an arbitrary element $s \in \overline{s(P)s(Q)}$ and a neighbourhood N of s. We can then find elements $s' \in s(P)$, $s'' \in s(Q)$ with small neighbourhoods N' and N'' such that $N'N'' \subset N$. But then $P \ast Q(N) \geqslant P \ast Q(N'N'') \geqslant P(N')Q(N'') > 0$, which implies $s \in s(P \ast Q)$, so that $\overline{s(P)s(Q)} \subset s(P \ast Q)$ as stated.

Note that if S is compact then the set $A = s(P)s(Q)$ is automatically closed and we can leave out the closure symbol.

Homogeneous stochastic processes could be introduced in several ways. The following one will sometimes be convenient but not quite the same as the one usually used on the real line (see Notes 2.2.$_4$).

Definition 2.2.3. By a homogeneous stochastic process on a semi-group S we mean a family of probability measures $\{P_t;\ 0 \leqslant t < \infty\}$ on S such that $P_t \ast P_s = P_{t+s}$, $0 \leqslant t$, $s < \infty$. It is said to be continuous (in probability) if $P_t \underset{\text{weakly}}{\to} \delta_e$ when $t \downarrow 0$.

Let us also define the *infinitely divisible distributions.*

Definition 2.2.4. A probability distribution $P \in \mathcal{P}(S)$ on a semi-group with unit element will be said to be infinitely divisible if for any natural number n there is a $P_n \in \mathcal{P}(S)$ such that

$$\begin{cases} P_n^{n*} = P \\ P_n \to \delta_e . \end{cases}$$

Remark. The distributions P_t appearing in a continuous homogeneous process are automatically infinitely divisible. There is no reason to believe that a given infinitely divisible distribution P can *always* be embedded in a homogeneous process P_t so that $P_1 = P$.

The condition $P_n \to \delta_e$ is essential but it may be of interest to see what happens when it is left out, especially of course on semi-groups with no unit element.

Note that we have used identical components P_n in the definition. If unequal components are allowed we may get a wider definition.

2.3. Compact stochastic semi-groups

For stochastic semi-groups as general as those described in the last section it does not seem possible at present to prove any substantial probabilistic results of the desired type. To do this it will be necessary to add more structure (say the group property or more) to S but before we do this in the following chapters we will study (without going into too much detail) what happens when S is compact.

Let us remember that the convolution operation $P_1 * P_2$ is continuous in P_1 and P_2. Also that the convergence used is weak* convergence (see Notes $2.3._1$) so that it follows that $\mathcal{P}(S)$ is compact (see Notes $2.1._3$). A compact semi-group has at least one idempotent (see Notes $2.3._2$) and we will find one idempotent below by an actual construction. The compactness of $\mathcal{P}(S)$ makes the search for limit theorems much more promising. Of course it can very well happen that a given sequence $P, P^{2*}, P^{3*}, \ldots$ does not converge; take e.g. a cyclic group with all the probability mass in an element different from the identity element. But we can compensate for such an oscillating behavior either by some operation involving the selection of a subsequence or by the application of a summability procedure. The latter alternative can be carried out as follows.

If $f(s) \in C(S)$, the set of continuous functions on S, with the usual maximum norm, then the *probability operator*

$$T f(s) = \int f(st) \, dP(t)$$

is a linear, bounded operator, $\|T\|=1$. Introduce the averaged sum of iterates

$$A_n = \frac{1}{n}(T + T^2 + \dots + T^n),$$

which is a probability operator with respect to the probability measure

$$\pi_n = \frac{1}{n}(P + P^{2*} + \dots + P^{n*}).$$

The family of functions $f(st)$ of the argument s and the parameter t is equicontinuous: for every $\varepsilon > 0$ and $s \in S$ there is a neighbourhood $N(s)$ such that

$$|f(s't) - f(st)| < \varepsilon \text{ for all } t \in S$$

if $s' \in N(s)$. This follows easily from the compactness of S using an indirect argument. But then the family of functions $A_n f(s)$ of the argument s and the parameter n is also equicontinuous. We can then find a convergent subsequence $A_{n_v} f(s)$, see Notes 2.3.$_3$. A version of the mean ergodic theorem (see Notes 2.3.$_4$) tells us that $A_n f(s)$ converges (in the norm of $C(S)$) to some $Af(s)$, and $AT = TA = A^2 = A$ and A is a positive linear transform in $C(S)$. Translated to probability language this means the following.

Theorem 2.3.1. The Cèsaro mean $\pi = \lim_{n \to \infty} 1/n(P + P^{2*} + \dots + P^{n*})$ exists and is an idempotent measure, $\pi^{2*} = \pi$. Further $P \ast \pi = \pi \ast P = \pi$.

It is possible to describe the possible limit measures and idempotents a bit more completely. First of all it is clear that any limit distribution will have its mass contained in the closure of the union of all the sets $[s(P)]^n$, $n = 1, 2, \dots$ Therefore we can just as well restrict ourselves to this subset of S; it is clearly also a compact topological semi-group. This will be assumed to have been done already. It can then be shown that $s(\pi)$ is the kernel K of S (see Notes 2.3.$_4$).

Now let P be any idempotent measure $P^{2*} = P$ with the support $s(P) = \Sigma$. Since $\Sigma = \Sigma \cdot \Sigma$ it is a subsemi-group. Let us denote its kernel by K. We are going to show that Σ must be completely simple (see Notes 2.3.$_2$). If $\Sigma = K$ this is clear since a kernel is completely simple. Otherwise $\Sigma - K$ is non-vacuous and it must contain some idempotent, since otherwise we would get a contradiction: if I is the set of all idempotents in Σ then $\Sigma = \Sigma I \Sigma$ (see Notes 2.3.$_2$), on the other hand all the idempotents would be in K so that $I \subset K$ and $\Sigma I \Sigma \subset \Sigma K \Sigma \subset K$ so that $\Sigma \subset K$ which contradicts the assumption. Let us denote by i such an idempotent in $\Sigma - K$ and let $f(s)$ be a continuous and non-negative function on Σ taking the value 1 at i and vanishing on K. Consider the function

$$f_1(s) = \int_S f(ts)\,P(dt).$$

Let s_0 be a point where $f_1(s)$ attains a maximum. Then

$$f_1(s_0) = \int_S f(ts_0)P(dt) = \int_S \int_S f(\tau\, ts_0)P(d\,\tau)P(dt) = \int_S f_1(ts_0)P(dt)$$

so that $f_1(s_0) = f_1(ts_0)$ for all t in the support $s(P) = \Sigma$. But since $Ks_0 \subset K$ it follows that the same maximum value is attained at some point k of K. However

$$f_1(k) = \int_S f(tk)P(dt) = 0$$

since $tk \in K$. The maximum value of f_1 is thus zero. On the other hand $f_1(i)$ must be positive since

$$f_1(i) = \int_S f(ti)P(dt) \geqslant \int_{N_i} f(ti)P(dt) \geqslant (1-\varepsilon)P(N_i) > 0$$

where N_i is some small neighbourhood of i. This is impossible which falsifies the assumption and proves that $\Sigma = K$, and since any kernel is a completely simple semi-group this proves the following

Theorem 2.3.2.a. If P is an idempotent probability measure on a compact semi-group then its support $s(P)$ is a completely simple subsemi-group.

It is possible to get a complete although somewhat indirect characterization of idempotent probability measures on completely simple semi-groups as follows. Such a semi-group can be represented as (see Notes 2.3.$_2$)

$$K = T \times X \times Y.$$

Here T is a compact topological group and X and Y are compact Hausdorff spaces. Multiplication by two elements

$$\begin{cases} s \;= (t,\ x,\ y\,) \\ s' = (t',x',y') \end{cases}$$

is defined through the relation

$$ss' = (t\varphi(x,y')t',x',y)$$

where φ is a continuous function taking values in T and defined on $X \times Y$. Then the following theorem holds.

Theorem 2.3.2.b. Let π be any (regular) idempotent measure on a compact semi-group. The support of π is a completely simple subsemi-group K and it can be written as

$$\pi = \tau \times \xi \times \eta$$

where τ is the normed Haar measure on T and ξ and η are regular probability measures on X and Y respectively (see Notes 2.3.$_2$).

If S is also commutative one can get more detailed information of the idempotent measures. The reason for this is that now the minimum ideal forms a group and of course this simplifies things.

Theorem 2.3.3. If S is a compact, commutative semi-group and P is an idempotent probability measure in $\mathcal{P}(S)$, then the support of P is a subgroup H of S and P is the normed Haar measure over H.

To prove this we note that we already know that the support H coincides with the kernel K. But K is now a group (see Notes 2.3.$_5$) and we shall see in Chapter 3, Theorem 3.2.1., that any idempotent probability measure defined on a compact group is a normed Haar measure.

We now easily get the following result.

Theorem 2.3.4. For a compact and commutative stochastic semi-group the averaged measures

$$\pi_n = \frac{1}{n}(P + P^{2*} + \ldots + P^{n*})$$

converge to the normalized Haar measure on the least ideal K of the semi-group spanned by the set $s(P)$.

Proof: We already know that the limit measure π of the sequence π_n exists, is idempotent and has as its support the minimal ideal K. The last theorem then tells us that π must be the unique normed Haar measure on K.

Now let us consider a homogeneous process of the discontinuous type

$$\begin{cases} P_t * P_u = P_{t+u}; \ t \text{ and } u \geqslant 0. \\ P_h(e) \to 1 \text{ as } h \downarrow 0 \end{cases}$$

on a compact, stochastic semi-group S with a unit element e. The associated probability operators

$$T_t f(s) = \int_S f(su) P_t(du), \ f \in C(S),$$

form a multiplicative semi-group. Since $T_h - I$ is the operator corresponding to the signed measure $M = P_h - \delta_e$ its norm is the absolute variation (see 2.1.)

$$\|T_h - I\| = \int_S |M(ds)| = 1 - P_h(e) + P_h(\bar{e}) = 2[1 - P_h(e)] \to 0$$

with h; the semi-group is continuous in the uniform operator topology. Then we know (see Notes 2.3.6) that the infinitesimal operator exists,

$$V = \lim_{h \downarrow 0} \frac{T_h - I}{h}$$

and that

$$T_t = \exp tV.$$

Apparently

$$Vf(s) = \int_S f(su) B(du)$$

where B is a bounded signed measure of variation 0 and no negative variation except at the point $s = e$. Going back from the operators to the measures we have proved the following

Theorem 2.3.5. Let P_t be a homogeneous process of discontinuous type on the compact, stochastic semi-group S. Then there exists a finite, signed measure B with variation 0, and the only possible place where negative variation is allowed is at the unit element e, and

$$P_t = \exp^* tB = \sum_{k=0}^{\infty} t^k \frac{B^{k*}}{k!}.$$

Since B can be written as $\lambda(Q - \delta_e)$ where $\lambda > 0$ and Q is a probability measure

$$P_t = \exp^* t\lambda(Q - \delta_e) = \exp(-t\lambda) \cdot \exp^* t\lambda Q,$$

$$= \sum_{k=0}^{\infty} \exp(-t\lambda) \frac{(t\lambda)^k}{k!} Q^{k*},$$

since δ_e and Q commute. This makes it possible to interpret P_t as a *compound Poisson distribution*: compounded of the given probability distribution Q and with a Poisson distribution of mean value λt. It is not our purpose to go into a detailed examination of the sample functions of the process, which would require the introduction of a separable version of the process, but it is clear that the above implies that they behave as in the classical case on the real line.

Instead of demanding that $P_h \to \delta_e$, in the sense specified, we now require that P_h tends to some measure I

$$\int_S |(P_h - I)\,(ds)| \to 0, \text{ as } h \downarrow 0.$$

It is then an easy modification of the previous proof to prove that

$$P_t = I \exp^* tQ,$$

where the notation remains unchanged, and I must be an idempotent measure and $Q = Q * I = I * Q$.

Now assume only that $P_h \to \delta_e$ in the usual weak sense. Then T_t is a continuous semi-group in the strong topology and $T_h f \to f$ strongly as $h \downarrow 0$ for any $f \in C(S)$. Hence we know from general semi-group theory that

$$T_t f = \lim_{\varepsilon \downarrow 0} \exp\, tV_\varepsilon f$$

strongly, with
$$V_\varepsilon = \frac{T_\varepsilon - I}{\varepsilon}.$$

Theorem 2.3.6. If P_t is a continuous, homogeneous process on a compact stochastic semi-group then P_t is the limit of compound Poisson processes

$$P_t = \lim_{\varepsilon \downarrow 0} \exp^* tB_\varepsilon$$

where
$$B_\varepsilon = \frac{P_\varepsilon - \delta_e}{\varepsilon}.$$

Note that if S is finite then the operators P_t (considered as operators on $C(S)$) are also continuous in the uniform operator topology. Hence the infinitesimal generator is bounded and defined throughout $C(S)$. Hence a continuous, homogeneous process on a finite semi-group must be a compound Poisson process.

The homogeneous processes are related to infinitely divisible laws and it seems appropriate to make some remarks about this here. To start with, suppose that P is a probability distribution such that for every natural number n there is a $P_n \in \mathcal{P}(G)$ with

$$\left.\begin{aligned} P &= P_n^{n*} \\ \sup_n\, &n[1 - P_n(e)] < \infty \end{aligned}\right\}.$$

For any rational $t = \nu/n$ we can define a homogeneous process $P_t, P_1 = P$ as will be done in section 3.2. To pass to irrational values of the parameter t it is sufficient (but not necessary) to show that $P_t(e) \to 1$ as $t \downarrow 0$ through rational values. But this is so because

$$P_t(e) = P_n^{\nu^*}(e) \geqslant [P_n(e)]^\nu \geqslant \left(1 - \frac{K}{n}\right)^\nu \to 1 \text{ as } \frac{\nu}{n} \to 0,$$

with $K = \sup\limits_n n[1 - P_n(e)]$. Hence the given distribution P has been embedded in a homogeneous process on the stochastic semi-group and what we know about such processes applies directly.

In this connection it is natural to prove the following simple limit law.

Theorem 2.3.6.a. Consider a sequence of probability measures $P_1, P_2, \ldots$ defined on the compact topological semi-group with the unit element $e \in S$, $P_n \in \mathcal{P}(S)$. As n increases the probability mass should be concentrated at e

$$P_n(e) = 1 - \frac{\lambda}{n} + o\left(\frac{1}{n}\right),$$

and the conditional probability distribution Q_n of s given $s \neq e$ should satisfy

$$\lim_{n \to \infty} \int_S |Q_n(ds) - Q(ds)| = 0$$

for some $Q \in \mathcal{P}(S)$. Then $P_n^{n^*}$ converges weakly to a compound Poisson distribution

$$\lim_{n \to \infty} P_n^{n^*} = \exp^* \lambda(Q - I) = e^{-\lambda} \sum_{\nu=0}^{\infty} \frac{\lambda^\nu}{\nu!} Q^{\nu^*}.$$

Especially if Q is concentrated at $s, Q = \delta_s$, then the limit is a Poisson distribution corresponding to a stochastic element of the form s_0^ν; ν is a Poisson variable with mean value λ.

Proof: Consider the probability operator

$$Tf(t) = \int_S f(ts) P(ds), \quad f \in C(S).$$

Then

$$Tf(t) = f(t)\left[1 - \frac{\lambda}{n} + o\left(\frac{1}{n}\right)\right] + \frac{\lambda}{n} \int_S f(ts) Q(ds) + \frac{\lambda}{n} \int_S f(ts)[Q_n(ds) - Q(ds)]$$

so that

$$T = I + \frac{\lambda}{n}(Q - I) + \Delta_n$$

where the symbol Q is now also used to denote the probability operator corresponding to the probability measure Q and where $\|\Delta_n\| = o(1/n)$. Since convolutions P^{n^*} correspond to powers T^n and since

$$T^n f = \left[I + \frac{\lambda}{n} (Q-I) + \Delta_n \right]^n f \to \exp^* \lambda (Q-I) f$$

the stated result follows.

The support of an arbitrary compound Poisson distribution on a compact semi-group can be characterized easily.

Proposition. A set $E \subset S$ is the support of some compound Poisson distribution if and only if E is a closed subsemi-group of S with a unit.

Proof: If P is a compound Poisson distribution

$$P = \exp^* \lambda (Q - \delta_e) = e^{-\lambda} \sum_{\nu=0}^{\infty} \frac{\lambda^\nu}{\nu!} Q^{\nu*},$$

so that $s(P) = \overline{\bigcup_0^\infty s(Q^{\nu*})}$ if $\lambda > 0$; the case $\lambda = 0$ is trivial. But P^{2*} has the same form but with the parameter 2λ so that $s(P^{2*}) = s(P)\,s(P) = s(P)$ which proves that $s(P)$ is a subsemi-group with a unit. As a support it is automatically closed.—If E is as stated then we can define a probability distribution Q with $s(Q) = E$ (we can choose a denumerable subset of Q everywhere dense in Q and assign positive probabilities to each element of the subset). Form $P = \exp^*(Q - \delta_e)$; it follows immediately that $s(Q^{\nu*}) = s(Q) = E$ and $s(P) = E$.—The proposition is not generally true for an arbitrary infinitely divisible distribution.

Now we take one step further in specialization and assume that S is a finite and commutative stochastic semi-group with elements $s_1, s_2, ..., s_n$. A probability distribution P over S is then characterized by discrete probabilities $p_1, p_2, ..., p_n$, $\Sigma p_\nu = 1$. We will also assume that there exists an integer m such that $s^{m+1} = s$ for all s (see Notes 2.3.$_7$).

Although this set up may seem special we shall give it some attention because this is where we first meet with *Fourier analysis*, this most important of all the analytical tools used in the study of probabilistic limit laws on general structures.

Definition 2.3.1. A complex valued function $\sigma(s) \not\equiv 0$ is called a semi-character if $\sigma(s)\sigma(t) = \sigma(st)$ for all $s, t \in S$. The set of all semi-characters is denoted by Σ.

The number of elements in Σ is the same as that of S, n. Σ is also a semi-group if and only if S has a unit as will be assumed. The n semi-characters are linearly independent.

Definition 2.3.2. By the Fourier transform $\hat{P}(\sigma)$ of P we mean the function defined on Σ

$$\hat{P}(\sigma) = E\,\sigma(s) = \sum_{\nu=1}^{n} p_\nu\,\sigma(s_\nu).$$

Theorem 2.3.7. a) P is uniquely defined by $\hat{P}$ through an inversion formula

$$p_\nu = \sum_\sigma a_\sigma\,\sigma(\nu), \quad \text{the uniqueness theorem.}$$

b) $\hat{P}_\nu \to \hat{P}$ is equivalent to $P_\nu \to P$, the continuity theorem.

c) $\widehat{(P_1 \ast P_2)} = \hat{P}_1 \hat{P}_2$ convolution corresponds to multiplication of the Fourier transforms.

d) A function $f(\sigma)$ can be represented as $\hat{P}$, the Fourier transform of a probability distribution over S if and only if $f(\sigma)$ is positive definite.

$$\sum_{\sigma,\,\psi\,\in\,\hat{S}} C_\sigma \bar{C}_\psi f(\sigma\bar{\psi}) \geqslant 0, \text{ and } f(1) = 1$$

where 1 denotes the semi-character that is identically equal to 1.

Proof: a) Since the semi-characters are linearly independent the determinant $\det(\sigma(s_\nu);\,\sigma(s_\mu)) \neq 0$ and we can solve the inhomogeneous system of n linear equations

$$\hat{P}(\sigma) = \sum_{1}^{n} p_\nu\,\sigma(s_\nu)$$

in the p's.

b) One direction of this statement is obvious, the other ones follow from the inversion formula.

c)
$$\widehat{(P_1 \ast P_2)} = \sum_{\nu,\,\mu=1}^{n} p_\nu\, p_\mu\, \sigma(s_\nu\, s_\mu)$$

$$= \sum_{\nu=1}^{n} p_\nu\, \sigma(s_\nu) \sum_{\mu=1}^{n} p_\mu\, \sigma(s_\mu) = \hat{P}_1(\sigma)\,\hat{P}_2(\sigma).$$

d) If $f(\sigma)$ is the Fourier transform of P we have $f(1) = \sum_\nu p_\nu \cdot 1 = 1$ and

$$\sum_{\sigma,\,\psi} C_\sigma \bar{C}_\psi f(\sigma\bar{\psi}) = \sum_{\substack{\sigma,\,\psi \\ \nu}} C_\sigma \bar{C}_\psi\, p_\nu\, \sigma\,(s_\nu)\, \bar{\psi}\,(s_\nu) = \sum_\nu p_\nu \left| \sum_\sigma C_\sigma\, \sigma(s_\nu) \right|^2 \geqslant 0. \tag{1}$$

On the other hand if $f(\sigma)$ is an arbitrary complex valued function on Σ it can be represented as the Fourier transform of some complex valued (not necessarily non-negative) function defined on S, say with values q_ν at s_ν. But choosing the c's so that

$$\sum_\sigma C_\sigma \, \sigma(s_\nu) = \begin{cases} 1 \text{ if } \nu = \mu \\ 0 \text{ if } \nu \neq \mu, \end{cases}$$

we see that q_μ is given by the expression on the left side of (1) and is non-negative. Also $\sum_\nu p_\nu = f(1) = 1$.

The main result on limit laws on commutative, finite stochastic semi-groups is the following.

Theorem 2.3.8. Let i be an idempotent in S such that $i \, s^m(P) \subset S_i = \{s \, | \, s^m = i\}$ and form the subgroup Σ_i spanned by $is^m(P)$. Then P^{n*} has a limit if and only if $is(P)$ has an element in common with Σ_i for any i defined above.

Let us just mention the basic idea of the proof. From what has been said above it is clear that the only situation when P^{n*} does not converge is when $\hat{P}(\sigma) = e^{i\theta}$ for some real $\theta \not\equiv 0 \bmod 2\pi$. But then

$$1 = |\hat{P}(\sigma)| \leqslant \sum_\nu p_\nu \, |\sigma(s_\nu)| \leqslant \sum_\nu p_\nu = 1$$

and we must have $\sigma(s_\nu) = e^{i\theta}$ for all $s_\nu \in s(P)$. It has been shown (see Notes 2.3.$_7$) that this is equivalent to the condition in the theorem. We will meet the same sort of reasoning later in the study of stochastic groups and we shall then go into more detail in the proofs.

Finite stochastic semi-groups (commutative or not) can be studied with advantage from the point of view of finite *Markov chains*. The transition probabilities are given by

$$p_{ij} = \sum_{s_i s_\nu = s_j} p_\nu \, ,$$

where the summation is over all ν such that $s_i s_\nu = s_j$, and we can now just borrow results from the theory of Markov chains. We know e.g. that if there is an integer k and a column index j such that $p_{ij}^{(k)} > 0$ for all i then P^{n*} converges to a limit distribution. If, in addition, S is also a group then

$$\sum_{i=1}^{n} p_{ij} = \sum_{s_\nu = s_i^{-1} s_j} p_\nu = \sum_\mu p_\mu = 1$$

where, in the second sum, j is kept fixed and i runs through all values between 1 and n. Hence the transition probability matrix is doubly stochastic. The limit distribution is then simply the uniform one $P(s = s_\nu) = 1/n$. In a similar way it is possible to treat more interesting cases with more zeroes in the transition probability matrix.

2.4. Illustrations

A few simple examples may be illuminating. Consider a stochastic semi-group of order two with the following multiplication table

	1	2
1	1	2
2	2	2

It is a commutative semi-group. The transition probability matrix is, in terms of the probabilities $p_1 = P(1)$ and $p_2 = P(2)$,

$$P = \begin{Bmatrix} p_1 & p_2 \\ 0 & 1 \end{Bmatrix}.$$

If p_1 and $p_2 > 0$ so that $s(P) = S$ then there is a limit distribution with $P(1) = 0$, $P(2) = 1$. Only if $p_1 = 1 - p_2 = 0$ can we get any positive limiting probability mass on 1 and then $P(1) = 1$.

With the same notation but with the multiplication table

	1	2
1	1	2
2	2	1

corresponding to a commutative group we get the transition probabilities

$$P = \begin{Bmatrix} p_1 & p_2 \\ p_2 & p_1 \end{Bmatrix}.$$

If p_1 and $p_2 > 0$ then there is the uniform limit distribution $P(1) = P(2)$ over the support of P. If $p_1 = 0$ then no limit distribution exists but the Césaro-mean of the P^{n*} is still the uniform distribution. If $p_1 = 1$ then there is the uniform limit distribution $P(1) = 1$ over the support of P.

Let S consist of the interval $[0, 1]$ with the binary operation $st = \max(s, t)$. For a given distribution P over $[0, 1]$ let M be the essential supremum of $s \in S$ with respect to this measure, $M = \sup s, P[0, s] < 1$. Of course it is enough to deal with the subsemi-group $[0, M]$. All the elements of S are idempotents and the kernel K of $[0, M]$ consists of the single element M.

From general considerations one knows that the Césaro mean of P^n* has all its mass on K; in this special case of course P^n* actually converges without any summability method to the degenerated limit measure with $P(M)=1$.

Let the stochastic variables $x_{n1}, x_{n2}, ..., x_{nn}$ be independent and all distributed according to the distribution function $F_n(x)$ on $[0,1]$. Then the (semi-group) product $x_{n1} x_{n2} ..., x_{nn}$ has the distribution function $F_n^n(x)$. If we assume that the limit in

$$G(x) = \lim_{n \to \infty} n[1 - F_n(x)]$$

exists in $[0,1]$ then $P_n^{n^*}$ tends to the limit distribution described by the function

$$F(x) = e^{-G(x)}$$

in $(0,1]$. $G(x)$ is a non-increasing function with $G(1)=0$ and $G(0) \leqslant +\infty$. It can be interpreted as the asymptotic expected number of values of x_{n1}, $x_{n2}, ..., x_{nn}$ falling into the interval $(x,1]$. Note the relation to the classical results in extreme value theory in probability.—What is the form of the continuous, homogeneous processes, P_t? Describing P_t through a distribution function $F_t(x)$ and since $F_t(x)$ satisfies

$$F_{t+s}(x) = F_t(x) \mathbin{\text{\Large$*$}} F_s(x) = F_t(x) F_s(x)$$

we see that $F_t(x)$ must have the form

$$F_t(x) = \exp - tG(x)$$

where $G(x)$ is a function with the same properties as above. It is easy to find the infinitesimal generator of $\{P_t\}$. Indeed, let $f(x) \in C'(S)$ where $C'(S)$ is the set of continuous functions on $S=[0,1]$ with

$$\int_0^1 |f(x) - f(0)| \, G(dx) \text{ convergent.}$$

This restricts the behavior of $f(x)$ only at $x=0$, and $C'(S)$ is everywhere dense in $C(S)$. Then we get

$$A_h f(x) = \frac{T_h - I}{h} f(x) = \frac{1}{h} \int_0^1 [f[\max \, (x,y)] - f(x)] P_h(dy)$$

$$= \frac{1}{h} \int_x^1 [f(y) - f(x)] F_h(dy) = - \int_x^1 [f(y) - f(x)] e^{-h \, G(y)} G(dy).$$

By dominated convergence this tends to

$$A f(x) = - \int_x^1 [f(y) - f(x)] G(dy)$$

when h tends to zero, so that the operator A with the domain $C'(S)$ is the infinitesimal operator. If, in particular, the value of $G(0)$ is finite then A is a bounded operator and P_t is a compound Poisson process

$$F_t(x) = \sum_{\nu=0}^{\infty} \frac{(t\lambda)^\nu}{\nu!} e^{-\lambda t} H^\nu(x)$$

where $H(x) = \lambda - G(x)$, $\lambda = G(0)$.—The question also arises if there are any normal processes on S in the following sense. Is there any continuous homogeneous process P_t with independent increments and with almost all the sample functions continuous? We deal only with separable processes. Consider a continuous homogeneous process x_t and let n be a large number. Then we can write

$$x_{\nu/n} = \max [\eta_1^{(n)}, \eta_2^{(n)}, \ldots, \eta_\nu^{(n)}], \quad 0 \leqslant \nu \leqslant n,$$

where the stochastic variables $\eta_\nu^{(n)}$ are independent and have the common distribution function

$$F_{1/n}(x) = \exp -\frac{1}{n} G(x).$$

Consider the points a and b, $0 < a < b < 1$, and such that $G(b) > 0$. Then the probability that the sequence $\eta_1^{(n)}, \eta_2^{(n)}, \ldots, \eta_n^{(n)}$ has $n-1$ values in $[0, a]$ and one value in $(b, 1]$ is

$$n \exp -\frac{n-1}{n} G(a) \left[1 - \exp -\frac{1}{n} G(b) \right].$$

For large values of n this is approximately equal to

$$p = G(b) \exp - G(a) > 0.$$

But this implies that with probability $p - \varepsilon$ some increment $x_{\nu+1/n} - x_{\nu/n}$ is larger than $b - a$. Hence the process is not continuous with probability one and there does not exist any normal process on this semi-group.

Defining instead $st = t, s$ and $t \in [0, 1]$, we get a non-commutative semi-group. Again all the elements are idempotents; the kernel is the whole semi-group. Since $P^{n*} = P$ a limit measure always exists trivially and it can have any form.

Another, perhaps less trivial, example consists of all the 2×2 matrices

$$A = \begin{Bmatrix} a_{11} & a_{12} \\ a_{21} & a_{22} \end{Bmatrix}$$

with $a_{ij} \geqslant 0$ and $\|A\| \leqslant 1$ where we use the norm

$$\|A\| = \max_i \sum_j |a_{ij}|,$$

with the usual property $\|AB\| \leqslant \|A\| \cdot \|B\|$. This is a compact semi-group. If $s(P) = S$ then $E\|A\| < 1$ (actually this holds already if there is positive mass on the set $\|A\| < 1$), and, for $B_n = A_1 A_2 \dots A_n$,

$$E\|B_n\| \leqslant [E\|A\|]^n \to 0$$

as n tends to infinity so that the product B_n of independent, stochastic semi-group elements converges to the zero matrix in probability. This is just what could be expected from the theory. Indeed since 0 is an idempotent element of S it is clear that the minimal ideal (the kernel of S)

$$K = \bigcap_i S i S, \quad i = \text{arbitrary indempotent},$$

must consist of the single element 0.

Let us consider the subset Σ of matrices of the form

$$A = \begin{Bmatrix} x & 1-x \\ y & 1-y \end{Bmatrix}$$

with $0 \leqslant x \leqslant y \leqslant 1$; it is a closed subsemi-group of the previous semi-group S. Let P have all its mass on some subset of Σ containing at least one point with x or $y \neq 0, 1$. Denoting

$$A_\nu = \begin{Bmatrix} x_\nu & 1-x_\nu \\ y_\nu & 1-y_\nu \end{Bmatrix}, \quad B_n = A_1 \dots A_n = \begin{Bmatrix} u_n & 1-u_n \\ v_n & 1-v_n \end{Bmatrix}$$

we get
$$\begin{aligned} u_{n+1} &= y_{n+1} + u_n(x_{n+1} - y_{n+1}) \\ v_{n+1} &= y_{n+1} + v_n(x_{n+1} - y_{n+1}) \end{aligned}$$

Iterating this it is not difficult to see that P^{n*} converges to a limit distribution given by the stochastic matrix

$$C = \begin{Bmatrix} z & 1-z \\ z & 1-z \end{Bmatrix}$$

where z is distributed like

$$z = y_1 + y_2(x_1 - y_1) + y_3(x_1 - y_1)(x_2 - y_2) + \ldots$$

where the (x_ν, y_ν) are independent for different values of ν and distributed according to P. The series converges almost certainly since, under the assumptions made, $E|x-y| < 1$ so that

$$\sum_1^\infty |E\, y_{k+1}(x_1 - y_1)(x_2 - y_2) \ldots (x_k - y_k)| \leqslant \sum_1^\infty [E|x-y|]^k < \infty.$$

This fits in well with the general theory. Indeed the idempotents are of the form

$$\begin{Bmatrix} 1 & 0 \\ 0 & 1 \end{Bmatrix} \quad \text{or} \quad \begin{Bmatrix} z & 1-z \\ z & 1-z \end{Bmatrix}$$

and using this we can determine the kernel as before and get after some computations

$$K = \text{the set of all matrices} \begin{Bmatrix} z & 1-z \\ z & 1-z \end{Bmatrix}, \quad 0 \leqslant z \leqslant 1.$$

In general there are also more degenerate limit distributions with all their mass in single idempotent elements of the form

$$\begin{Bmatrix} 1 & 0 \\ 1 & 0 \end{Bmatrix}, \quad \begin{Bmatrix} 0 & 1 \\ 0 & 1 \end{Bmatrix} \quad \text{or} \quad \begin{Bmatrix} 1 & 0 \\ 0 & 1 \end{Bmatrix}$$

It is also possible to study denumerable stochastic semi-groups by direct application of the theory of Markov chains with denumerably many states.

Let $P_t = \exp^* t(Q - \delta_e)$ and $P'_t = \exp^* t(Q' - \delta_e)$ be two compound Poisson processes on a compact semi-group S. If S is commutative it is obvious that $R_t = P_{t*} P'_t$ is also a compound Poisson process $= \exp^* t(Q + Q' - 2\delta_e)$. The same holds on an arbitrary compact semi-group if Q and Q' (or P_t and P'_t) commute.

What happens if S is not necessarily commutative but we know that R_t is a compound Poisson process? Putting $R_t = \exp^* tR$ we have by direct calculation $R = Q + Q' - 2\delta_e$.

Expanding $\exp^*$ into powerseries and comparing the second order terms we get

$$(Q - \delta_e) * (Q' - \delta_e) = (Q' - \delta_e) * (Q - \delta_e).$$

Choose $Q = \delta_s$, $Q' = \delta_{s'}$, we find that the above relation reduces to $ss' = s's$. In other words: if the convolution of any two compound Poisson processes always results in a compound Poisson process then the semi-group S must be commutative.

Recalling the role the compound Poisson processes play for the limit theorems we see that when we study limit theorems for convolutions of non-identical probability measures we cannot expect to get as simple extension of the theory as on the real line. On the real line the limit theorems are quite insensitive to whether the components are identical or not. On a non-commutative semi-group the situation is a good deal more complicated.

Here we have discussed this difficulty in a special context, but the same phenomenon occurs throughout the whole theory of non-commutative stochastic structures. In this book we shall not enter into a detailed discussion of the case with unequal components.

CHAPTER 3

—

STOCHASTIC GROUPS;

COMPACT AND COMMUTATIVE CASES

3.1. Generalities on stochastic groups

In this and the following two chapters we shall deal with locally compact and separable groups (see Notes 2.1). The group property will make it possible to arrive at more detailed results on the probabilistic properties of the structures.

A classical result on topological groups, in its original form due to Haar, tells us that there is a non-trivial left *invariant*, (regular), Borel measure μ so that for any Borel set $E \subset G$ and element $g \in G$ one has $\mu(gE) = \mu(E)$ and $P(0) > 0$ for any open set 0. This measure, the Haar measure, is uniquely determined except for multiplication by a real constant. Of course there is also a similar right invariant measure ν.

For an arbitrarily fixed g the set function $\mu(Eg)$ is again left invariant and can then be written as $\Delta(g) \cdot \mu(E)$. It is almost immediate that $\Delta(gh) = \Delta(g)\Delta(h)$ and that $\Delta(g)$ is a continuous positive function on G. If G is compact $\Delta(g)$ must be equal to 1, because if there were an element g_0 with $\Delta(g_0) \neq 1$, say >1, then the sequence $\Delta(g_0^n) = \Delta^n(g_0)$ is unbounded although $\Delta(x)$ must be bounded because of its continuity and the compactness of its domain. Hence left and right invariant measures are essentially the same on a compact group (on a commutative group the same holds true trivially) and we can norm them both in the same way $\mu(G) = \nu(G) = 1$, since compact sets have finite measures.

The set $\mathcal{D}(G)$ of all (regular), normed Borel measures on G is a topological semi-group just as before. One may think that $\mathcal{D}(G)$ could now be a group, but this is never so except when G consists only of e. Indeed let a given probability distribution $P \in \mathcal{D}(G)$ have the inverse $Q, P * Q = \delta_e$, where δ_g means the probability distribution with all its mass at the element g. But then $\overline{s(P)s(Q)} = e$, and this is only possible if $s(P)$ consists of a single point.

An important subset of $\mathcal{D}(G)$ is the set $\mathcal{A}(G)$ of all the distributions in $\mathcal{D}(G)$ which are absolutely continuous with respect to μ (see Notes 3.1$_1$). They can be written as

$$P(E) = \int_E p(g)\mu(dg),$$

where $p(g)$ is the generalized *frequency function* or the Radon-Nikodym derivative $dP/d\mu$. If Q is an arbitrary distribution in $\mathcal{D}(G)$ then $Q*P \in \mathcal{A}(G)$ with the generalized frequency function

$$\frac{d(Q*P)}{d\mu} = \int_{h \in G} p(h^{-1}g)Q(dh)$$

so that $\mathcal{A}(G)$ is a left ideal. If Q also belongs to $\mathcal{A}(G)$ with $dQ/d\mu = q$ then

$$\frac{d(Q*P)}{d\mu} = \int_{h \in G} p(h^{-1}g)q(h)\mu(dh).$$

Related to $\mathcal{A}(G)$ is the group algebra $L_1(G)$ consisting of all complex valued functions absolutely integrable with respect to Haar measure over G. If multiplication in $L_1(G)$ is defined by a convolution integral as above it is clear that $L_1(G)$ is a Banach algebra (see Notes 3.1.$_2$).

In the same way we define the Hilbert space $L_2(G)$ as the set of all complex valued functions quadratically integrable with respect to ν and with the usual L_2-definition of inner product.

If $f \in L(G)$ the probability operator T is defined through a stochastic group element h with a probability distribution P

$$Tf(g) = Ef(gh) = \int_{h \in G} f(gh)P(dh).$$

But the domain of T can be extended to $L_2(G)$. Indeed we have, with $L_2(G)$-norms (see Notes 3.1.$_3$),

$$\|Tf\|^2 \leqslant \int_G \int_G \int_G |f(gh_1)f(gh_2)|P(dh_1)P(dh_2)\nu(dg)$$

$$\leqslant \int_G \int_G \sqrt{\int_G |f(gh_1)|^2\nu(dg)}\sqrt{\int_G |f(gh_2)|^2\nu(dg)}\,P(dh_1)P(dh_2)$$

and

$$\int_G |f(gh)|^2\nu(dg) = \int_G |f(u)|^2\nu(du\,h^{-1}) = \int_G |f(u)|^2\nu(du) = \|f\|^2$$

so that $\|Tf\| \leqslant \|f\|$, $f \in L$, which ensures the unique extension to $L_2(G)$ in the customary manner since $L(G)$ is dense in $L_2(G)$.

A special case of considerable interest is the operator

$$R_h f(g) = f(gh)$$

corresponding to the probability measure $P = \delta_h$. R_h is called a *right translation*. The left translations could be defined as

$$L_h f(g) = f(h^{-1}g).$$

The adjoint of T is

$$T^* f(g) = \int_{h \in G} f(gh) \, Q(dh),$$

where $Q(E) = P(E^{-1})$ so that $T^* f(g) = E f(gh^{-1})$ with the same notation as above. Indeed $(Tf, \varphi) = \int_G \int_G f(gh) \, \overline{\varphi(g)} \, P(dh) \, \nu(dg) = \int_G \int_G f(u) \, \overline{\varphi(uv)}$ $Q(dv) \nu(du) = (f, T^* \varphi)$; f and $\varphi \in L(G)$ and hence also in $L_2(G)$. Necessary and sufficient for T to be self-adjoint is that $P(E) = Q(E) = P(E^{-1})$, P should be a *symmetric measure*. In order that T be a normal operator it is necessary and sufficient that $P * Q = Q * P$; this is shown in the same way.

The symmetric case is of special interest. The spectrum of T is then situated in the real interval $(-1, 1)$, and it is of interest to see if $\|T\| = 1$ occurs or if $\|T\| < 1$. Inequality implies that $T^n \to 0$ as $n \to \infty$ so that $P^{n^*}(C) \to 0$ for any compact set C. This can of course not happen if G is compact. Let us assume that for any given compact C we can find a sequence $\varphi_n(g) \in L(G)$ such that the constant function 1 is approximated uniformly on C by

$$\psi_n^{(h)} = \frac{\varphi_n * \tilde{\varphi}_n}{\|\varphi\|^2} = \frac{1}{\|\varphi\|^2} \int_G \varphi_n(g) \, \bar{\varphi}_n(gh) \, \nu(dg),$$

where $\tilde{\varphi}_n(g) = \bar{\varphi}_n(g^{-1})$. (We will meet a similar condition later on in Chapter 5.) Then

$$\frac{\|T \varphi_n\|^2}{\|\varphi_n\|^2} = \frac{1}{\|\varphi_n\|^2} \iiint \varphi_n(gh_1) \, \bar{\varphi}_n(gh_2) \, P(dh_1) P(dh_2) \nu(dg)$$

$$= \iint \psi_n(h_1^{-1} h_2) \, P(dh_1) P(dh_2)$$

which tends to 1 as n tends to infinity so that $\|T\|$ must be equal to 1.

Let us make some preliminary comments about T. Writing T in its spectral representation as a self-adjoint operator

$$T = \int_{-1}^{1} \lambda \, dE(\lambda),$$

where $E(\lambda)$ is a resolution of the identity, we have for the transform corresponding to P^{2n*}

$$T^{2n}f = \int_{-1}^{1} \lambda^{2n} d\,E(\lambda)f \rightarrow (E_1 + E_{-1})f = Sf$$

as n tends to infinity. Here E_1 and E_{-1} are the jumps of $E(\lambda)$ at $\lambda = \pm 1$ respectively. A limit measure of P^{2n*} (which is not a priori necessarily normed to one) must be an idempotent measure since $S^2 = S$. If $\lambda = \pm 1$ are not eigen-values then $T^n f \rightarrow 0$ and we can again conclude that all the probability mass escapes to infinity.

Suppose that P is a distribution with mass in the neighborhood of the unit element, $e \in s(P)$. If T has the eigen-value 1 there exists a function $f \in L_2(G)$ such that

$$f(g) = \int f(gh) P(dh).$$

In order that this be possible all the functions $f(gh)$, when $h \in s(P)$, must be equal for almost all g. It may be helpful for the reader to consider geometrically these functions as vectors in $L_2(G)$. Introducing the set

$$H = \{h \,|\, f(gh) = f(g) \text{ for almost all } g\}$$

we see that $s(P)$ must be contained in H. On the other hand H is obviously a closed subgroup of G and must then coincide with G. As usual we assume that we have chosen G already as the smallest closed subgroup (of the original group) containing $s(P)$. But then the eigen-function is a constant a.e. over G and such a function can belong to $L_2(G)$ only if G is compact. On the other hand, if G is compact then $f(g) \equiv 1$ is an eigen-function corresponding to the eigen-value $\lambda = 1$.

This links up with what was said above about symmetric distributions. Indeed if P is a symmetric measure then the sequence $T^{2n}f$ always converges, so that P^{2n*} converges vaguely to some measure P^{∞}. The limit measure is not necessarily normed; it may even vanish, $P^{\infty}(E) = 0$ for any compact set E. This depends upon the existence of the possible eigenvalues $\lambda = \pm 1$ of T, i.e. $\lambda = 1$ of T^2. But T^2 corresponds to P^{2*} which measure always has the unit element e in its support, since P was assumed to be symmetric.

Let us now prove the following result valid under the standard conditions (separable, locally compact group) and where we do not assume any symmetry property for P.

Theorem 3.0. Given a probability measure $P \in \mathcal{P}(G)$, form the averages π_n of the iterates $P^{\nu *}$

$$\pi_n = \frac{1}{n} \sum_{\nu=1}^{n} P^{\nu *}.$$

Then π_n converges vaguely to some limiting measure π with $\pi(G) \leqslant 1$. Either π is an idempotent probability measure, $\pi^{2*} = \pi$, $\pi(G) = 1$, or $\pi = 0$, the zero measure.

Proof: Starting from the probability operator

$$Tf(g) = \int_G f(gh) P(dh)$$

we can use ergodic theory (see Notes 3.1.$_4$) to prove that

$$f_n = T_n f = \frac{1}{n} \sum_{\nu=1}^{n} T^\nu f = f_n$$

converge when n tends to infinity. It will be adequate to let $f \in L(G)$ as will be assumed below.

We now claim that for any positive ε there is a neighborhood N_ε of e such that

$$|f(g'h) - f(gh)| \leqslant \varepsilon \text{ for all } h \in G$$

if $g'g^{-1} \in N_\varepsilon$. To prove this let C be a compact set outside of which $f(g)$ vanishes and choose a neighborhood M of e. The set $\overline{M}C$ is compact and we can find a neighborhood N_ε such that N_ε and $N_\varepsilon^{-1} \subset M$ and

$$|f(x') - f(x)| < \varepsilon$$

if x' and $x \in \overline{M}C$ and if $x'x^{-1} \in N_\varepsilon$. We suppose that $g'g^{-1} \in N_\varepsilon$ and separate four cases. If 1) $g'h$ and $gh \in \overline{M}C$ and

$$d = |f(g'h) - f(gh)|$$

then $d \leqslant \varepsilon$. If 2) $g'h$ and gh are both outside of $\overline{M}C$ then $f(g'h) = f(gh) = 0$ and $d = 0$. If 3) $gh \notin \overline{M}C$ then $g'h \notin C$ since otherwise $gh = g(g')^{-1}g'h \in MC$ contrary to assumption; hence $d = 0$. Similarly if 4) $g'h \notin \overline{M}C$.

The equicontinuity property that we have proved for the functions $f(gh), h \in G$, also holds for the functions $T^\nu f$ and the functions

$$T_n f = \frac{1}{n} \sum_{\nu=1}^{n} T^\nu f = f_n.$$

The functions $f_n(g)$ converge almost certainly to some function $\varphi(g)$ as n tends to infinity. Using the equicontinuity it follows that $f_n(e) \to \varphi(e)$ so that

$$\lim_{n \to \infty} \int_G f(h)\pi_n(dh) = \varphi(e).$$

But $\varphi(e) = Sf$ is a positive linear functional in $L(G)$ and can be written as

$$Sf = \int_G f(h)\pi(dh),$$

where π is a measure with $\pi(G) \leqslant 1$. It is associated with an operator R, say in $L_2(G)$,

$$Rf(g) = \int_G f(gh)\pi(dh).$$

The operator R is idempotent $R^2 = R$. If $\|R\| < 1$ the equation $R(R - I) = 0$ implies $R = 0$ and $\pi(G) = 0$. On the other hand if $\|R\| = 1$ then we must have $\pi(G) = 1$. Otherwise, with $\pi(G) = k$, $0 < k < 1$, the operator R/k would be a probability operator. Such an operator has a norm $\leqslant 1$ which would be a contradiction. This completes the proof of the theorem.

This theorem has the following corollary.

Corollary. If the support of P is contained in a compact subsemi-group of G we have $\pi(G) = 1$ and π_n converges weakly to π.—If $\pi(G) = 1$ the support of P is contained in a compact subgroup of G.

Proof: The first part is obvious.— On the other hand if $\pi(G) = 1$ there must be a function $f \in L_2(G)$ such that $f = Tf$. We can now reason as above to show that $s(P)$ is contained in a compact group.

The reader who is puzzled by the relation between the two statements in the corollary may see the connection clearer if he observes that a compact subsemi-group of a group forms a subgroup, since it obeys a cancellation law.

In the classical theory for addition of independent, stochastic variables the main analytical tool is Fourier analysis. To extend this to stochastic groups one could try to introduce a transform

$$\hat{P}(r) = \int_G r(g)\,dP(g), \quad r \in R,$$

where R is a set of functions $r(g)$ on G taking values in some as yet unspecified space V. We would like to have just as in the classical situations

$$\widehat{P_1 * P_2}(r) = \hat{P}_1(r) \cdot \hat{P}_2(r).$$

Obviously we must have addition, scalar multiplication and multiplication defined in V. Taking $P_1 = \delta_{g_1}$, $P_2 = \delta_{g_2}$ we get

$$r(g_1 g_2) = r(g_1) \cdot r(g_2)$$

and this leads us to consider the group representation of G. Also it is necessary that V is wide enough so that we have uniqueness between distribution and transform, $P \leftrightarrow \hat{P}$. It turns out that in the commutative case we can take V as the set of complex numbers $e^{i\theta}$ of modulus one; in the compact case the right choice of V is the set of all finite dimensional unitary matrices, while for a general locally compact group we must use the set of unitary transformations of a Hilbert space. The last case is considerably more complex than the two first ones and not at all as well penetrated. For didactic reasons we shall therefore treat the compact and commutative groups in the following two sections while the discussion of the general locally compact groups will be postponed till Chapter 5.

Before turning to the Fourier analysis let us deal briefly with the following problem. Consider a homomorphic mapping h of G on G' where G and G' are locally compact groups. See Notes $3.1._5$.

Theorem 3.1. A probability distribution $P \in \mathcal{P}(G)$ induces one $P' \in \mathcal{P}(G')$ through the definition

$$\int_G f(g') P'(dg') = \int_G f[h(g)] P(dg)$$

for any $f \in L(G')$. If $P_n \in \mathcal{P}(G)$ and $P_n \to P$ weakly then $P'_n \to P'$ weakly. If P_t is a homogeneous process on G such that $\lim_{t \downarrow 0} P_t(NK) = 1$, where N is an arbitrary neighborhood of e and K is the kernel of the homomorphism, then P'_t is a continuous, homogeneous process on G'.

Proof: Consider the linear functional

$$I(f) = \int_G f[h(g)] P(dg).$$

Note that h must be an open continuous mapping. If $f(g')$ is bounded and continuous on G' then $f[h(g)]$ has the same property on G and $I(f)$ defines uniquely a regular probability measure P' on G' such that $I(f) = \int_{G'} f(g') P'(dg')$. For any open set $0 \subset G$ the set $0' = h(0) \subset G'$ is open and $P(0) = P'(0')$.

Now assume that $P_n \to P$ weakly. Then it follows directly that $P'_n \to P'$

weakly.—Actually it would have been enough to assume that $P_n \to P$ weakly modulo K; the reader can give this statement a precise meaning and proof without difficulty.

Now let P_t be a homogeneous process. We have for any bounded and continuous $f(g')$

$$\int_G f(g') P'_{t+s}(dg') = \int_G f[h(g)] P_{t+s}(dg) = \int_G \int_G f[h(xy)] P_t(dx) P_s(dy)$$

$$= \int_G \int_G f[h(x) h(y)] P_t(dx) P_s(dy) = \int_{G'} \int_{G'} f(x' y') P_t(dx') P_s(dy')$$

which implies $P'_{t+s} = P'_t \ast P'_s$ so that P'_t is a homogeneous process. To see that P'_t is also weakly continuous let f be continuous and bounded and consider

$$\int_{G'} f(g') P'_t(dg') = \int_G f[h(g)] P_t(dg) \to f(e') = \int_{G'} f(g') \delta_{e'}(dg')$$

since the probability mass of P_t is assumed to converge to the kernel K of the homomorphism and $h(g) = e'$ when $g \in K$. Hence $P'_t \to \delta_{e'}$ weakly as stated.

If a group G has been given a probability structure with distributions, homogeneous processes and limit results, then theorem 3.1. implies the analog for the factor group $F = G/N$, if N is a closed, normal subgroup. We just use the natural homomorphism of G on F.

A word should be said about the definition of homogeneous stochastic processes on groups. To get a definition more in accordance with customary practice on the real line than Definition 2.2.3 we could say that a homogeneous process on a group is a stochastic process $g(t)$, $0 \leqslant t < \infty$, taking values in G and such that the stochastic elements $g^{-1}(t_\nu) g(t_{\nu+1})$, $0 < t_1 < t_2 < \dots$, are independent (this could be called left invariance and of course we could also use right invariance) and the probability distribution of $g^{-1}(t) g(s)$ depends only upon the difference $s - t$. To this would be added continuity in probability. If we also have almost all sample functions continuous it would be natural to speak of a *Brownian motion* on the group. The study of the individual sample functions is however outside the scope of this book.

3.2. Compact stochastic groups

Let us first study the idempotent measures on a compact group G. They are completely characterized through the following theorem.

Theorem 3.2.1. If $P \in \mathcal{D}(G)$ is an idempotent measure on a compact group G then $s(P)$ is a closed subgroup H of G; P is the normed Haar measure on H.

Proof: Compare the proof used in 2.3. for the corresponding theorem there. We can use the same idea taking advantage of the fact that now $H = s(P)$ satisfies $H^2 = H$ so that H is a subgroup (see Notes 2.3.$_2$). Then the ideal $I = H$ and the function g takes its supremum on the whole H, i.e. g is a constant and P is the invariant measure on H.

Let us now turn to Fourier analysis of probability distributions on compact groups. First we shall recapitulate the basic facts on unitary representations of such groups. By a unitary representation of degree n we mean a function $M(g)$ taking as values unitary $n \times n$ matrices, $M(g) = \{m_{ij}(g); \ i,j = 1, 2, \ldots n\}$, such that $m_{ij}(g)$ are continuous functions on G and satisfy the fundamental relation $M(g_1) M(g_2) = M(g_1 g_2)$ for all $g_1, g_2 \in G$. Such a representation is called irreducible if there is no subspace of unitary n-space left invariant by $M(g)$ for all $g \in G$. Instead of representations one ought really to speak of classes of equivalent representations: two representations $M_1(g)$ and $M_2(g)$ are said to be equivalent if $M_1(g) = A M_2(g) A^{-1}$, $g \in G$. From each equivalence class we will select one single representative of the class.

When we go through the set R of all irreducible, non-equivalent unitary representations r, the various components $m_{ij}^{(r)}(g)$ give us a set of functions on G with interesting properties. In the Hilbert space $L_2(G)$ we have the orthogonality relation $m_{ij}^{(r)}(g) \perp m_{kl}^{(s)}(g)$ if $r \neq s$. Further the n^2 elements of the matrix $M^{(r)}(g)$ are orthogonal to each other and of norm $n^{-1/2}$. For the separable groups that we are working with $L_2(G)$ is separable and hence R is denumerable so that we can enumerate the representations by $r = 0, 1, 2, \ldots$, where we will reserve the subscript $r = 0$ for the identity representation $M(g) \equiv I$.

A fundamental result, the Peter-Weyl theorem, tells us that the set of all the functions $m_{ij}^{(r)}(g)$ is complete in $L_2(G)$. It is even uniformly complete, i.e. every continuous function on G can be uniformly approximated by finite linear combinations of the $m_{ij}^{(r)}(g)$. A consequence of this is that for every $g \neq e$ there is an r such that $M^{(r)}(g) \neq I$, or in words, the irreducible representations are sufficiently many to be able to separate the elements of the group.

These considerations lead naturally to the

Definition 3.2.1. By the Fourier transform of a probability measure P on the compact group G we mean the sequence of matrices

$$\hat{P}_r = EM^{(r)}(g) = \int_G M^{(r)}(g) P(dg), \quad r = 0, 1, 2, \ldots$$

The integral should of course be interpreted as integration element-wise of the matrices. We now study the fundamental properties of this sort of Fourier transform.

Theorem 3.2.2. On a compact group the Fourier transform of a probability distribution behaves as follows.

a) The probability distribution P is uniquely determined by $\hat{P}_r, r = 0, 1, 2, \ldots$

b) Consider a sequence of probability distributions $P, P^{(1)}, P^{(2)}, \ldots$ and their Fourier transforms $\hat{P}_r, \hat{P}_r^{(1)}, \hat{P}_r^{(2)}, \ldots$ Then $P^{(n)} \to P$ is equivalent to $\hat{P}_r^{(n)} \to \hat{P}_r$, for all $r = 0, 1, 2, \ldots$

c) Convolution of two probability measures corresponds to multiplication of the Fourier transforms, $\widehat{P^{(1)} * P^{(2)}} = \hat{P}^{(1)} \cdot \hat{P}^{(2)}$.

d) For any r the matrix $\hat{P}_r$ represents a bounded linear transformation of bound $\|\hat{P}_r\| \leqslant 1$. For the identity representation, $r = 0$, we always have $\hat{P}_0 = I$.

Proof: For a given continuous function $f(g)$ on G we can approximate it uniformly by finite linear combinations of the elements $m_{ij}^{(r)}(g)$. If $\hat{P}^{(1)} = \hat{P}^{(2)}$ it then follows that

$$\int_G f(g) P_1(dg) = \int_G f(g) P_2(dg)$$

for every continuous function $f(g)$, which proves a).

In the same way if $\hat{P}^{(n)} \to \hat{P}$ it follows for every continuous $f(g)$ that

$$\lim_{n \to \infty} \int_G f(g) P_n(dg) = \int_G f(g) P(dg),$$

i.e. $P_n \to P$. The direct part of b), that $P_n \to P$ implies $\hat{P}_n \to \hat{P}$ is obvious. The mapping $P \to \hat{P}$ is a homeomorphism.

To prove c) it is sufficient to observe that

$$\widehat{P^{(1)} * P^{(2)}} = \int_G \int_G M(g_1 g_2) P^{(1)}(dg_1) P^{(2)}(dg_2)$$

$$= \int_G M(g_1) P^{(1)}(dg_1) \int_G M(g_2) P^{(2)}(dg_2) = \hat{P}^{(1)} \cdot \hat{P}^{(2)}.$$

The mapping $P \to \hat{P}$ is an isomorphism.

The last statement of the theorem, d), follows from $\|M_r(g)\| = 1$, $g \in G$, since P is a probability measure. For $r = 0$ the representation $M_r(g)$ is simply the identity I, and $\hat{P}_0 = EI = I$.

For a given set of matrices $A_0, A_1, A_2, ...$, we would like to know if there is a $P \in \mathcal{D}(G)$ such that $\hat{P}_r = A_r, r = 0, 1, 2, ...$ and also in that case construct P from the A's: inversion of the Fourier transform. First the simplest case, if the series

$$p(g) = \sum_{r,i,j} n_r a_{ij}^{(r)} \bar{m}_{ij}^{(r)}(g); \quad \{a_{ij}^{(r)}\} = A_r;$$

converges uniformly to a non-negative function, and if $A_0 = I$, then A_r is the Fourier transform of the absolutely continuous probability distribution with frequency function $p(g)$. To see this note that $p(g)$ is continuous and hence integrable over $G, p \in L_1(G)$. Further, from the orthogonality relations for the m's,

$$\int_G m_{ij}^{(r)}(g) p(g) \mu(dg) = a_{ij}^{(r)},$$

and especially for $r = 0$

$$\int_G p(g) \mu(dg) = 1,$$

which proves the assertion. In the general situation we must apply some summability method for the inversion. One such method is as follows. Consider a decreasing sequence of neighborhoods $N_1 \supset N_2 \supset N_3 ...$ of the unit element converging to e, $\bigcap_k N_k = e$. For every N in this sequence let $f^N(g)$ be a non-negative function, vanishing outside of N and with

$$\int_G f^N(g) \mu(dg) = 1.$$

These functions can be chosen with a great deal of freedom except that they should be regular enough so that the series

$$\sum_{r,i,j} n_r f_{rij}^N, \quad f_{rij}^N = \int_G f^N(g) m_{ij}^{(r)}(g) \mu(dg),$$

should converge absolutely.

Theorem 3.2.3. If for any $N = N_k$ the series

$$p_k(g) = \sum_{r,i,j,\alpha} n_r a_{i\alpha}^{(r)} f_{r\alpha j}^N \bar{m}_{ij}^{(r)}(g),$$

which converges uniformly, is non-negative and $A_0 = I$, then $\{A_r\}$ is the Fourier transform of a $P \in \mathcal{D}(G)$. P is the weak limit of the absolutely continuous distributions with frequency functions $p_k(g)$.

Proof: Since $|m_{ij}^{(r)}(g)| \leqslant 1, |a_{ij}^{(r)}| \leqslant 1$, the series converges uniformly, and we have just seen that then the matrices

$$\left\{ \sum_\alpha a_{i\alpha}^{(r)} f_{r\alpha j}^N; \quad i,j = 1, 2, \dots n_r \right\} = A^r F_r^N,$$

where
$$F_r^N = \{f_{rij}^N; \quad i,j = 1, 2, \dots n_r\},$$

are the Fourier transforms of the absolutely continuous distributions with frequency functions $p_k(g)$. But for any r we have $F_r^N \to I$ as $N \downarrow e$, so that $A^r F_r^N \to A_r$ and the proof is complete.

One can formulate a number of other criteria in order that a given set of matrices $A^{(r)}$ should be Fourier transform of a probability distribution on G (see Notes 3.2.$_1$). However, just as on the real line these criteria are a bit awkward to use in concrete situations.

The study of the behavior of P^{n*} for large values of n is reduced, through the isomorphism $P \to \hat{P}$, to the study of $(\hat{P})^n$. But high matrix powers depend strongly upon the largest (in absolute value) eigen-value λ of the matrix, and it is important to find criteria ensuring $|\lambda| < 1$.

Lemma 3.2.1. If some $\hat{P}_r, r \neq 0$, has an eigen-value λ on the unit circle, $|\lambda| = 1$, then the support $s(P)$ must be contained in a closed proper subgroup A of G or in a coset $g_0 A$ of A.

Proof: Suppose that $\lambda = e^{i\alpha}$ is an eigen-value of $\hat{P}_r$ so that there is a nontrivial vector z such that $\hat{P}_r z = \lambda z$. Then

$$\|z\| = \|\hat{P}_r z\| = \left\| \int_G M^{(r)}(g) P(dg) z \right\| \leqslant \int_G \|M^{(r)}(g) z\| P(dg) \leqslant \|z\|$$

and equality is possible if and only if

$$s(P) \subset A_\alpha = \{g \mid M^{(r)}(g) z = e^{i\alpha} z\}.$$

The set $A_0 = \{g \mid M^{(r)}(g) z = z\}$ is a closed, proper (note that the $M^{(r)}(g)$ are irreducible) subgroup. The observation that $A_\alpha = g_0 A_0, g_0 \in A_\alpha$, completes the proof.

It can actually be shown that if $s(P)$ is not contained in any such subgroup or coset then $\|\hat{P}_r\| < 1$, see Notes 3.2.$_1$.

If the probability distribution is not concentrated on any such subgroup or coset, then it would follow that $\widehat{P_r^{n*}} = (\hat{P}_r)^n \to 0$ as $n \to \infty$ for every $r \neq 0$. What is the corresponding limit distribution?

Lemma 3.2.2. The uniquely determined probability distribution P with the Fourier transform

$$\dot{\hat{P}}_0 = I; \ \hat{P}_r = 0, \ r = 1, 2, \ldots;$$

is the normalized Haar measure on G.

Proof: To see that $\hat{P}_r$ vanishes for the Haar measure and $r \neq 0$ it is enough to notice that the elements of $\hat{P}_r$ can be considered as inner products in $L_2(G)$

$$\{\hat{P}_r\}_{ij} = \int_G m_{ij}^{(r)}(g) P(dg) = (m_{ij}^{(r)}, \ 1)$$

of $m_{ij}^{(r)}(g)$ and the constant function 1. The statement then follows from the orthogonality relations for all the functions $m_{ij}^{(s)}(g)$.

If $e \in s(P)$ then $s(P)$ cannot be contained in a coset of any proper closed subgroup. This gives us the following partial result.

Lemma 3.2.3. If $s(P)$ contains the unit element e then P^{n*} converges to the normalized Haar measure on G or on one of its closed subgroups.

A more complete limit result on compact stochastic groups is the following.

Theorem 3.2.4. For a given probability distribution P the limit of P^{n*}, $n \to \infty$, exists if and only if $s(P)$ is not contained in any coset of any closed, proper, normal subgroup of G. The limit of P^{n*} is normalized Haar measure on G.

Proof: As usual we assume that G has been modified, if necessary, to be the smallest closed subgroup containing $s(P)$. The "if" part of the proof is similar to what has just been proved. Indeed if the limit did not exist it can be seen (e.g. by representing $\hat{P}_r$ in its Jordan canonical form; the reader may prefer to argue via the norm of P_r as in Notes 3.2.$_1$) that some $\hat{P}_r$ must have an eigen-value $\lambda = e^{i\alpha}$ of modulus one, so that we know that $s(P)$ would be contained in a closed, proper subgroup A_0 of G if $\alpha = 0$, which case has been excluded. If $\alpha \not\equiv 0 \bmod 2\pi$, then $s(P) \subset g_0 A_0 = A_0 g_0$. Introduce the closed set $B = \{g \mid gA_0 = A_0 g\}$. But B is a closed subgroup containing $g_0 A_0$ and $s(P)$ so that we must have $B = G$, i.e. A_0 is a normal subgroup.

To prove the "only if" part let us consider the stochastic product element $\gamma_m = g_1 g_2 \ldots g_m$ where g_ν is of the form $g_0 m$; m is a stochastic element in the closed, proper, normal subgroup N of G and $g_0 \notin N$. Because of the normal property we can rearrange the factors and see that γ_n is of the form $g_0^n k_n$ where the stochastic element k_n takes values in N. For $n = 1, 2, 3, \ldots$ we will get elements in the sets

$$g_0 N, \ g_0^2 N, \ g_0^3 N, \ \ldots$$

and it is intuitively clear that this is incompatible with convergence. A complete proof runs as follows. Introduce the closed, commutative subgroup $C = \{g_0^n \,|\, n = 0, \pm 1, \pm 2, \ldots\}$. The set CN is a closed subgroup of G (since N is normal) and as $s(P) \subset CN$ we must have $CN = G$. Introduce the homomorphism $g \to gN$ of C on to the commutative (and compact) group G/N. We will now use the character χ on G/N and we know that there exists a character (use the completeness property) χ such that χ is 1 on N and $e^{i\alpha} \neq 1$ on $g_0 N$. Extend the character to $k(g)$ on G by $k(g) = \chi(gN)$, so that $k(g) = \chi(g_0 N) = e^{i\alpha}$ for $g \in g_0 N$. Further we have the Fourier transform

$$\int_G k(g) \, P(dg) = \int_{s(P)} k(g) \, P(dg) = e^{i\alpha},$$

since $s(P) \subset g_0 N$. But we cannot get convergence of $e^{i\alpha n}$ as n tends to infinity so that P^{n^*} does not converge contrary to our assumption, and the proof is complete (see Notes 3.2._1).

It is easy to see that the limit of P^{n^*}, when it exists, is the normalized Haar measure.

This makes it possible to decide questions of convergence on compact stochastic groups. A case of some interest is when $s(P)$ contains the unit element; then P^{n^*} converges. Or if P is symmetric, then $s(P^{2^*})$ contains the unit element, so that P^{2n^*} converges.

The question naturally arises if the P^{n^*} get closer to some $m \in \mathcal{D}(G)$ in some monotonic manner as n increases.

Theorem 3.2.5. Introduce the measure of deviation between two distributions P and Q

$$d(P, Q) = \sup_E |P(E) - Q(E)|.$$

If m is such a distribution that $m * P = m$, then

$$d(P^{(n+1)^*}, m) \leqslant d(P^{n^*}, m).$$

Proof: We have

$$\left.\begin{aligned}
P^{(n+1)^*}(E) &= \int_G P^{n^*}(Ey^{-1}) \, P(dy) \\[2mm]
m(E) &= \int_G m(Ey^{-1}) \, P(dy)
\end{aligned}\right\}$$

using an idea that goes back to Markov we get

$$|P^{(n+1)^*}(E) - m(E)| \leqslant \int_G |P^{n^*}(Ey^{-1}) - m(Ey^{-1})| \, P(dy) \leqslant d(P^{n^*}, m)$$

which proves the inequality. The usefulness of the criterion d is limited. If for example the P_n's are singular with respect to m then $d_n \equiv 1$ and does not tell us anything of interest.—We can say something about the sort of m's that can satisfy $m * P = m$. Take the Fourier transform of this relation, $\hat{m}_r(\hat{P}_r - I) = 0$. If all the matrices $\hat{P}_r - I$ for $r \neq 0$ are non-singular, then $\hat{m}_r = 0$ for $r \neq 0$, so that m is Haar measure on G. Otherwise, for some $r \neq 0$, the matrix $\hat{P}_r - I$ must have a vanishing eigenvalue, i.e. $\hat{P}_r$ must have the eigenvalue 1. But we know that this is only possible in the degenerate situation that P has all its mass on a closed proper subgroup of G.—The criterion d has an especially simple geometric interpretation for absolutely continuous distributions. Say that we want to compare the Haar measure μ with $Q \in \mathcal{A}(G)$, $Q(dg)/\mu(dg) = q(x)$. Then

$$Q(E) - \mu(E) = \int_E [q(g) - 1] \mu(dg)$$

so that
$$d(Q, \mu) = \sup_E |Q(E) - \mu(E)| = \int_F [q(g) - 1] \mu(dg),$$

where $F = \{g \,|\, q(g) > 1\}$. In other words, the deviation d between Q and μ is the volume contained between the frequency surfaces described by the functions 1 and $q(g)$ in this order.

The following has some bearing on the same problem. Let us assume that $P(E) \geqslant cm(E)$, where m is normalized Haar measure on the group, E is an arbitrary Borel set and c is a constant, $0 < c < 1$. This implies that $s(P) = G$ and that the Radon-Nikodym derivative $m(dg)/P(dg)$ exists and is bounded. Introduce the measure $Q = P - cm = D + (1 - c)m$, where D is the signed measure $D = P - m$. Then

$$Q^{n*} = D^{n*} + (1 - c)^n m$$

since the convolutions $D * m = P * m - m = 0$ and $m * D$ vanish. This implies for any Borel set E that

$$P^{n*}(E) = D^{n*}(E) + m(E) = Q^{n*}(E) + [1 - (1 - c)^n] m(E)$$

so that
$$P^{n*}(E) - m(E) \geqslant -(1 - c)^n m(E) \geqslant -(1 - c)^n.$$

Taking the complementary set we get instead

$$P^{n*}(E) - m(E) \leqslant (1 - c)^n.$$

This proves

Theorem 3.2.6. If

$$P(E) \geqslant cm(E)$$

for all Borel sets E and for some constant c, $0 < c < 1$, then

$$\left| P^{n*}(E) - m(E) \right| \leqslant (1-c)^n.$$

(See Notes 3.2.3.)

We can study homogeneous processes on G by comparing with what was said in section 2.3.

This implies that a given set of matrix valued functions Q_t is the Fourier transform $\hat{P}_t$ of a continuous homogeneous process if and only if the following holds. On the set of functions $\mathcal{U}$ consisting of all finite linear combinations of $u_{ij}(g)$ there is defined a functional L which is the limit of functionals L_n of the form

$$L_n f(g) = \sum_{\nu=1}^{n} c_\nu^{(n)}[f(g_\nu^{(n)}) - f(e)], \quad c_\nu^{(n)} \geqslant 0$$

(the meaning of this limit should be that $L_n u \to Lu$ for every $u \in \mathcal{U}$). Q_t should be representable as

$$Q_t = \exp tA,$$

where the matrix $A = \{a_{ij}\}$ is given by its elements

$$a_{ij} = Lu_{ij}(g).$$

We now turn to the infinitely divisible distributions on G. We want a criterion for a given probability distribution to be embeddable in a continuous, homogeneous process.

Theorem 3.2.7. A probability distribution $P \in \mathcal{P}(G)$ is said to be embeddable in a continuous stochastic process if there is such a process Q_t and $P = Q_1$. P is embeddable in a continuous, homogeneous process if and only if there are $P_n \in \mathcal{P}(G)$, $n = 1, 2, \ldots$, such that

$$P = P_n^{n*}$$

with

$$\|\hat{P}_n - I\| = 0\left(\frac{1}{n}\right).$$

Proof: The "only if" part is almost immediate. Indeed, if $P = Q_1$ we can decompose P into factors $Q_{1/n}, P = Q_{1/n}^{n*}$. Consider the continuous matrix valued function Q_t. Since the matrices are finite dimensional this function is also continuous in the uniform operator topology and the semi-group property $\hat{Q}_{t+s} = \hat{Q}_t \hat{Q}_s$ implies the existence of a bounded infinitesimal generator A. But then

$$\lim_{n\to\infty} n(\hat{Q}_{1/n} - I) = A$$

so that
$$\|\hat{Q}_{1/n} - I\| \sim \frac{1}{n}\|A\| = 0\left(\frac{1}{n}\right) \quad \text{as stated.}$$

Assume instead that P can be decomposed into factors P_n which are small in the sense that $\|\hat{P}_n - I\| = 0(1/n)$. We are going to define Q_t for $t \geqslant 0$. Let $r_1, r_2, \ldots$ be the non-negative rational numbers. Consider the probability measures $P_n^{[nr_\nu]*}$. We can find a subsequence $n_1, n_2, \ldots$ such that the weak limits

$$\lim_{k\to\infty} P_{n_k}^{[n_k r_\nu]*} = Q_{r_\nu}$$

exist for $\nu = 1, 2, \ldots$; just use the compactness of $\mathcal{P}(G)$ and the diagonal selection procedure. Note that $Q_1 = P$. If r and s are non-negative rational numbers

$$P_{n_k}^{[n_k(r+s)]*} = P_{n_k}^{[n_k r]*} * P_{n_k}^{[n_k s]*} * P_{n_k}^{\varepsilon*},$$

where ε can take only the values 0 and 1. Taking limits we get

$$Q_{r+s} = Q_r * Q_s.$$

To extend this definition to irrational values of the parameter it is enough to show that $Q_r \to \delta_e$ when $r \to 0$ through rational, non-negative numbers. But we have

$$\|\hat{Q}_r - I\| = \lim_{k\to\infty} \|\hat{P}_{n_k}^{[n_k r]} - I\|$$

and
$$\|\hat{P}_{n_k}^{[n_k r]} - I\| \leqslant \|\hat{P}_{n_k} - I\| \sum_{0}^{[n_k r]-1} \|\hat{P}_{n_k}^{\nu}\| \leqslant r n_k \|\hat{P}_{n_k} - I\| = 0(r)$$

and this completes the proof.

It is not known at present if the bound $0(1/n)$ can be replaced by $o(1)$. The latter bound is equivalent to $P_n \to \delta_e$ weakly and on the real line this is known to be sufficient. It can also be shown to be sufficient if the P_n are symmetric distributions; just study the eigenvalues of the Hermitian matrices $\hat{P}_n$. In the general compact case this reasoning does not seem to work, the trouble arising from the fact that the nth roots are multiple valued functions in the complex plane. Or put in other terms, we have difficulty in separating the branches of the logarithmic function.—It may be mentioned that if P is infinitely divisible, then $\hat{P}$ is a non-singular matrix.

Remark. If we demand more, that the bound $0(1/n)$ should hold also for the norms $\|P_n - \delta_e\|$ then we see that Q_t becomes a compound Poisson process.

It is possible to express the spectrum of an operator

$$Tf(g) = \int_G f(gh)\,P(dh)$$

in terms of the spectra of the matrices $\hat{P}_r$. Let us do this for a symmetric probability distribution.

Theorem 3.2.8. The spectrum of the operator T corresponding to the symmetric probability distribution P is discrete. It consists of the eigenvalues of all the matrices $\hat{P}_r$ where an eigenvalue for $\hat{P}_r$ should be counted d_r times if $\hat{P}_r$ is a $d_r \times d_r$ matrix.

Proof: Let λ be an eigen-value of $\hat{P}_r$. Since $\hat{P}_r$ is Hermitian it can be diagonalized and the value λ will appear in the diagonal. In other words we choose a suitable new rectangular coordinate system. Let us call the functions $m_{ij}^{(r)}(g)$ expressed in the new coordinates $n_{ij}^{(r)}(g)$. Of course, the new functions $n_{ij}^{(r)}(g)$ have similar properties as the old ones, orthogonality etc. We then have

$$\left\{ \int_G n_{ij}^{(r)}(g)\,P(dg);\ i,j = 1, 2, \ldots d_r \right\} = \begin{Bmatrix} a_1 & & 0 \\ & a_2 & \\ & & \ddots \\ 0 & & a_{d_r} \end{Bmatrix},$$

where e.g. $a_1 = \lambda$. Then

$$T n_{k1}(g) = \int_G n_{k1}(gh)\,P(dh) = \sum_{j=1}^{d_r} n_{kj}(g) \int_G n_{j1}(h)\,P(dh) = a_1 n_{k1}(g).$$

This implies that $n_{k1}(g)$ is an eigen-function of T for $k = 1, 2, \ldots d_r$. But then λ is an eigen-value of T and should be counted d_r times. Since the $n_{ij}^{(r)}(g)$ form a complete system just as the $m_{ij}^{(r)}(g)$ do, it follows that all such eigenvalues exhaust the spectrum of T.

3.3. Commutative locally compact stochastic groups

Turning to groups which are commutative but not necessarily compact, only locally compact, the irreducible, unitary transformations now take the following form. They are no longer necessarily denumerable, but this complication is balanced by the fact that they are one-dimensional, $n_r \equiv 1$. Such a *character* $\gamma = (g, \gamma)$ is a continuous, complex valued function on G satisfying the equations

$$\left. \begin{aligned} (g + h, \gamma) &= (g, \gamma) \cdot (h, \gamma) \\ |(g, \gamma)| &\equiv 1 \end{aligned} \right\}.$$

The binary group operation is now written as addition. Introduce the *dual group* Γ consisting of all characters and define addition by

$$(g, \gamma + \delta) = (g, \gamma) \cdot (g, \delta),$$

and with neighborhoods of the form

$$\{\gamma \mid |1 - (g, \gamma)| < \varepsilon\}$$

for all g's in a compact set C and for a positive ε. It is known that Γ is locally compact and commutative and has G as its dual group. A discrete G corresponds to a compact dual group and a compact G has a discrete dual. For any $g \neq 0$ there is a $\gamma \in \Gamma$ such that $(g, \gamma) \neq 1$.

For any $P \in \mathcal{P}(G)$ we define the Fourier transform of P as the complex valued function on Γ

$$\hat{P} = \hat{P}(\gamma) = \int_G (g, \gamma) P(dg), \gamma \in \Gamma.$$

Theorem 3.3.1. On a commutative, locally compact group the Fourier transform $\hat{P}$ has the following properties:

a) P is uniquely determined by $\hat{P}$.

b') Consider a sequence of probability distributions $P_1, P_2, \ldots$ converging to $P \in \mathcal{P}(G)$. Then $\hat{P}_n \rightarrow \hat{P}$.

b") Consider a sequence of probability distributions $P_1, P_2, \ldots$ with Fourier transforms $\hat{P}_n(\gamma)$ converging to a continuous function $\hat{P}(\gamma), \gamma \in \Gamma$. Then $\hat{P}$ is the Fourier transform of a probability distribution P and $P_n \rightarrow P$ weakly.

c) $|\hat{P}(\gamma)| \leq 1$.

d) $\widehat{P_1 * P_2} = \hat{P}_1 \cdot \hat{P}_2$.

e) $\hat{P}(\gamma)$ is continuous.

Proof: The proof of the statements a), b'), c) and d) is similar to that of the corresponding statements in 3.2. We will base the proof of b") on Theorem 3.3.2, which tells us that the functions $\hat{P}_n(\gamma)$ are positive definite. The limiting continuous function $\hat{P}(\gamma)$ is then also positive definite and must be the Fourier transform of some $P \in \mathcal{P}(G)$. It is known (see Notes 5.1) that we can approximate the constant 1 uniformly on every compact set C by continuous positive definite functions p, vanishing outside of compact sets and taking the value 1 at 0. Now let P_{n_ν} be a subsequence converging vaguely to some measure $Q, Q(G) \leq 1$. We have to show that $Q(G) = 1$. Since p is positive definite it can be written as

$$p(g) = \int_\Gamma (g,\gamma)\mu(d\gamma), \mu \in \mathcal{P}(\Gamma).$$

Hence

$$\int_\Gamma p(g) P_{n_\nu}(dg) = \int_G \int_\Gamma (g,\gamma)\,\mu(d\gamma) P_{n_\nu}(dg)$$

$$= \int_\Gamma \hat{P}_{n_\nu}(\gamma)\mu(d\gamma) \to \int_\Gamma \hat{P}(\gamma)\mu(d\gamma) = \int_G p(g) P(dg),$$

and we also have

$$\int_G p(g) P_{n_\nu}(dg) \to \int_G p(g) Q(dg).$$

But now we take C very large and it follows that $Q(G)=P(G)=1$. The rest of the proof of b″) is standard.

It is easy to prove statement e). Indeed for any $\varepsilon > 0$ we can choose a compact set C with $P(C^*) < \varepsilon$ so that

$$|\hat{P}(\gamma) - \hat{P}(\gamma')| \leqslant 2\varepsilon + \int_C |(g,\gamma) - (g,\gamma')| P(dg)$$

which can be made small if γ' is in a small neighborhood of γ.

An important relation between positive definite functions and Fourier transforms of probability distributions is given by the following celebrated theorem that we state without proof (see Notes 3.3).

Theorem 3.3.2. If $p(\gamma)$ is a continuous, positive definite function with $p(0)=1$, then it is the Fourier transform $\hat{\mu}$ of some $\mu \in \mathcal{P}(G)$,

$$p(\gamma) = \int_\Gamma (g,\gamma)\mu(dg),$$

and vice versa.

We can now proceed to the study of idempotent probability distributions. Suppose that $P \in \mathcal{P}(G)$ is an idempotent, $P^{2^*}=P$, and let us as usual assume that G has been chosen as the smallest closed subgroup containing $s(P)$. If the Fourier transform $\hat{P}(\gamma)$ were a constant on the cosets of some non-trivial closed subgroup $\Gamma_0 \subset \Gamma$

$$\int_G (g,\gamma+\delta) P(dg) = \int_G (g,\gamma) \cdot (g,\delta) P(dg) = \int_G (g,\gamma) P(dg), \gamma \in \Gamma, \delta \in \Gamma_0,$$

then $(g,\delta)P(dg)=P(dg)$, $\delta \in \Gamma_0$. Hence $(g,\delta)=1$ for g in $s(P)$ so that this support would be contained in the annihilator sub-group $\{g \,|\, (g,\delta)=1 \text{ for } \delta \in \Gamma_0\}$

of Γ_0 and this has been excluded. Defining the new (complex) measure $m_\alpha, \alpha \in \Gamma$, by $m_\alpha(dg) = (g, \alpha) P(dg)$ it follows that $\hat{m}_\alpha(\gamma) = \hat{P}(\gamma + \alpha) \not\equiv \hat{P}(\gamma)$ if $\alpha \neq 0$ so that $m_\alpha \neq P$ if $\alpha \neq 0$. Now choose a compact subset C of G so big that $P(C^*) < \frac{1}{4}$. Introducing the open subset V of Γ

$$V = \{\gamma \mid |1 - (g, \gamma)| < \tfrac{1}{3}, g \in C\}$$

we have for $\alpha \in V$

$$\|P - m_\alpha\| = \int_G |1 - (g, \alpha)| P(dg) \leqslant \tfrac{1}{3} P(C) + 2 P(C^*) < \tfrac{5}{6} < 1.$$

On the other hand, $\|P - m_\alpha\| \geqslant \sup_\gamma |\hat{P}(\gamma) - \hat{m}_\alpha(\gamma)| = \sup_\gamma |\hat{P}(\gamma) - \hat{P}(\gamma + \alpha)| \geqslant 1$ except if $\alpha = 0$ since $\hat{P}$ takes only the values 0 and 1 as $(\hat{P})^2 = \hat{P}$. This shows that V contains only the element 0 so that Γ is discrete and hence G is compact. This proves the following theorem (see Notes 3.3.$_1$) which reduces the study of idempotent measures back to the compact case.

Theorem 3.3.3. An idempotent probability measure on a commutative, locally compact group has all its mass concentrated on a compact subgroup.

It is not difficult to modify Lemma 3.2.1. to the present situation. Indeed, if $\hat{P}(\gamma) = e^{i\alpha}$ all the mass of P must be in the set $A_\alpha = \{g \mid (g, \gamma) = e^{i\alpha}\}$ which is a coset of the closed subgroup A_0.

Lemma 3.3.1. A probability measure P on a locally compact commutative group whose Fourier transform $\hat{P}(\gamma)$ takes a value $e^{i\alpha}$ of modulus one must have all its mass on a coset A_α of a closed subgroup A_0 of G. Then $\hat{P}(\gamma) = e^{i\alpha}$ for all γ in a coset of a closed sub-group B of Γ annihilating $A_\alpha, (g, \gamma) = 1$ for $g \in A_\alpha, \gamma \in B$.

On a locally compact (but not compact) commutative group Haar measure is infinite and can not be normed to a probability measure. Therefore Theorem 3.2.5. does not apply directly. It is still true, in a certain sense, that *convolutions tend to flatten out distributions* over G. Indeed let us consider the class $\mathcal{A}(G)$ of absolutely continuous distributions and define a *measure of concentration* by

$$C(P) = \int_G p^2(g) \mu(dg),$$

where $p(g) = \dfrac{P(dg)}{\mu(dg)}$. Consider a distribution P_1 of finite concentration $C(P_1)$ and convolve it with P_2. The concentration of $P_1 * P_2$ is then

$$C(P_1 \ast P_2) = \int_G \left[\frac{P_1 \ast P_2(dg)}{\mu(dg)} \right]^2 \mu(dg)$$

$$= \int_\Gamma |\hat{P}_1(\gamma)\hat{P}_2(\gamma)|^2 m(d\gamma) \leqslant \int_\Gamma |\hat{P}_1(\gamma)|^2 m(d\gamma)$$

$$= \int_G \left[\frac{P_1(dg)}{\mu(dg)} \right]^2 \mu(dg) = C(P_1)$$

from Parseval's relation; $m(d\gamma)$ stands for the appropriately normed Haar measure on the character group Γ.

Theorem 3.3.4. With the above definition of concentration the convolution operation does not increase the concentration (see Notes $3.3._2$).

As we have seen the Fourier analysis of probability distributions on locally compact, commutative groups is quite similar to what we are used to on the real line etc. It is the more remarkable that our present knowledge of the probabilistic limit theorems on such groups is not more complete than it is to-day. See Notes $3.3._4$.

In this connection one would of course ask for a measure of dispersion behaving more or less like the variance. An immediate generalization of the variance is possible in a linear space, see 6.1., but we can arrive at some interesting concepts also in the absence of a linear structure.

Say at first that G is a locally compact, commutative group. For a given probability distribution P on G form the quantity

$$d_\gamma(P) = -\log|\hat{P}(\gamma)|,$$

which is a non-negative real number and the value $+\infty$ is allowed. For any character $\gamma \in \Gamma$ the functional $d_\gamma(P)$, defined in $\mathcal{P}(G)$, is additive

$$d_\gamma(P_1 \ast P_2) = -\log|\widehat{P_1 \ast P_2}| = -\log|\hat{P}_1| - \log|\hat{P}_2| = d_\gamma(P_1) + d_\gamma(P_2).$$

If we operate on these d_γ (when γ varies in the dual group Γ) in some linear way we will arrive at some additive functional $d(P)$ that we will take as our measure of dispersion $d(P) = \mathcal{L}d_\gamma(P)$.

We have not yet specified in what sense the linear operator $\mathcal{L}$ should be understood. We may do it via a Borel measure H defined on Γ

$$d(P) = \mathcal{L}d_\gamma(P) = \int_\Gamma d_\gamma(P)H(d\gamma).$$

It is convenient to assume that the variation of H is spread out over the whole group Γ, $s(H) = \Gamma$. We then have

Theorem 3.3.5. The functional

$$d(P) = -\int_\Gamma \log |\hat{P}(\gamma)| H(d\gamma),$$

is a measure of dispersion in the sense that

a) $0 \leqslant d(P) \leqslant \infty$
b) $d(P_1 * P_2) = d(P_1) + d(P_2)$
c) $d(P) = 0$ if and only if P reduces to a degenerate distribution δ_g
d) the operation of convolution does not decrease $d(P)$.

The proof of this theorem is quite simple. The statements a), b) and d) are immediately verified. To see that c) holds we observe that $d(P) = 0$ implies that

$$|d_\gamma(P)| = 0$$

for all $\gamma \in \Gamma$, since the support of H was the whole of Γ and $d_\gamma(P)$ is a continuous function of γ. Form the symmetrized probability measure $S = P * \bar{P}$ corresponding to the stochastic elements gh^{-1} (where g and h are independent and have the distribution P). The Fourier transform of S is $\hat{P}(\bar{P}) = |\hat{P}|^2 = 1$ so that S is the degenerate distribution δ_e. Hence P must be concentrated at some constant element g which proves the assertion. The converse is obvious.

It may be of interest to consider other forms of the operator $\mathcal{L}$, perhaps especially those that are local at the unit element e. By this we mean that the operator involves only the $d_\gamma(P)$ where γ is close to e. We shall not pursue this any further, but the reader will note that this is related to the way in which the ordinary variance is introduced. Another problem that we have not studied is what are the properties of the "convergence" defined through the relation $d(P_n) \to 0$.

If G is instead a compact group, the above must be modified since a Fourier transform P now consists of a sequence of matrices $\hat{P}_0, \hat{P}_1, \ldots$ If we replace the absolute value in the definition of $d_\gamma(P)$ by the norm we are led to a measure of dispersion of the form

$$d(P) = -\sum_{n=0}^{\infty} H_n \log \|\hat{P}_n\|,$$

where the H_ν are positive constants. This functional is no longer necessarily additive but since

$$\|\widehat{P_n^{(1)} * P_n^{(2)}}\| = \|\hat{P}_n^{(1)} \hat{P}_n^{(2)}\| \leqslant \|\hat{P}_n^{(1)}\| \cdot \|\hat{P}_n^{(2)}\|,$$

it follows that
$$d(P_1 * P_2) \geqslant d(P_1) + d(P_2);$$

the functional is superadditive. Note that the normed Haar measure has infinite dispersion just as would be expected.

Finally, if G is a general locally compact group, we would have to start from Fourier transforms of the type used in Chapter 5. We have not investigated the measures of dispersion one then gets.

In case G is a commutative group with an invariant metric $(g',g'') = (g'+h,g''+h)$ we may introduce a measure of dispersion $d(P)$ as follows (if G is a Lie group we could think of (g',g'') given by an invariant Riemann metric). We can write (g',g'') in the additive notation as $\|g'-g''\|$. Note that this is not a linear norm since $c \cdot g$ does not necessarily have a meaning for a scalar c. Define.

$$d(P) = \inf_{g \in G} \left[\int_G \|g-h\|^2 P(dh) \right]^{\frac{1}{2}}.$$

This quantity is always well defined if the value $+\infty$ is admitted. We cannot claim in general that there is a unique value g that realizes the minimum; think of a compact group with normed Haar measure. We have $0 \leqslant d(P) \leqslant +\infty$ and $d(P)=0$ if and only if P is degenerated to a single element of G. Further

$$d(P' * P'') \leqslant d(P') + d(P'').$$

To see this consider with $g = g' + g''$

$$\|h'+h''-g\| \leqslant \|h'-g'\| + \|h''-g''\|.$$

Here we think of h' and h'' as two independent stochastic group elements with the probability distributions P' and P'' respectively. Squaring and integrating we get the stated relation using Schwarz' inequality. Note that we do not in general have equality: a typical example is G = compact group and $P' = P'' =$ normed Haar measure

$$d(P' * P'') = d(P') < 2d(P').$$

(See also Notes 3.3.$_3$.)

In a similar way as in section 3.2. we can state and prove

Theorem 3.3.6. Let P be a symmetric probability distribution on a locally compact, commutative group G. The probability operator

$$Tf(g) = \int_G f(gh) P(dh)$$

has as its spectrum the set of values (Wertevorrat) of the Fourier transform $\hat{P}(\gamma)$.

The proof is left to the reader.

3.4. Illustrations

Consider the cyclic group G of m elements $0, 1, 2, \ldots m-1$ with group operation defined as addition modulo m. The characters are $(g, \gamma) = e^{2\pi i \frac{g\gamma}{m}}$, $\gamma = 0, 1, 2, \ldots m-1$ forming the group Γ. A probability distribution P over G with the probabilities $P(g=\nu) = p_\nu$, $\nu = 0, 1, 2, \ldots m-1$, has the Fourier transform

$$\hat{P}(\gamma) = \sum_{\nu=0}^{m-1} p_\nu \, e^{2\pi i \nu\gamma/m}$$

The nth convolution, P^{n*}, with the Fourier transform $[\hat{P}(\gamma)]^n$ obviously converges to the uniform distribution, with the Fourier transform $= 1$ for $\gamma = 0$ and zero otherwise if (and only if) $|\hat{P}(\gamma)| < 1$ for $\gamma \neq 0$. Hence we get this convergence if the support of P is not contained in the coset of any subgroup of G, and of course any subgroup of G is a cyclic group of an order which is a divisor of m.

The simple properties of the Fourier analysis of $\mathcal{P}(G)$ make it possible to examine this stochastic group in some detail.

The circle group T^1 is almost as simple. Let $G = T^1 = \{g \mid 0 \leqslant g < 2\pi\}$ with ordinary addition and identification modulo 2π. The characters are now $e^{i\gamma g}$, $\gamma = \ldots -1, 0, 1, \ldots$ so that we get the Fourier transform

$$\hat{P}(\gamma) = \int_0^{2\pi} e^{i\gamma g} \, P(dg).$$

Again the P^{n*} converge to the uniform distribution if P is not contained in the coset of any subgroup. This settles the convergence problem completely since the subgroups are just finite cyclic groups, and direct inspection of $s(P)$ will answer the problem.

We leave to the reader to examine these two cases in the light of the general discussion in 3.2. and 3.3. Related groups such as T^r can be studied without any additional trouble.

As far as finite groups are concerned the infinitely divisible distributions can be characterized in the following way (see Notes 3.4). Let G be a finite group of order n and let H be an arbitrary subgroup of G. Denote by μ_H the normed uniform distribution over H. Consider the group algebra A so that $f \cdot g$ means convolution of the two functions f and h defined on G.

Theorem 3.4.1. Choose an element $f(g) \in A$ such that

  1) $f(g)$ is H-invariant: $fh = hf = f$, $h \in H$.

  2) $\sum\limits_{g} f(g) = 0$.

  3) $f(g) \geqslant 0$ for $g \notin H$

and define
$$P = \exp{}_H f = \mu_H + \sum_1^\infty \frac{f^\nu}{\nu!} .$$

These and only these P can be decomposed $P = P_n^{n*}$, $n = 1, 2, \ldots$

If we add the condition $P_n \to \delta_e$ we get the representation of P

$$P = \exp f$$

as a compound Poisson distribution; here $H = \{e\}$.

CHAPTER 4

STOCHASTIC LIE GROUPS

4.1. Preliminaries on Lie groups

In a differentiable manifold we might hope to be able to introduce a probability calculus and to describe its probability distributions in terms of derivatives, differential equations etc. (see Notes $4.1._1$). However, in accordance with the main theme of this book we will limit ourselves to (Lie) groups. To avoid obscuring the main ideas by complicated, even if essentially simple, approximation arguments we shall discuss mainly the basic steps in the derivations and only sketch how the remaining details can be dealt with. The reader who wants a more complete discussion should consult the papers referred to in the Notes.

First we need some terminology and notation. Let G be a separable Lie group of dimension d and denote by $\Lambda(G)$ the corresponding *Lie algebra of infinitesimal right translations* of G. Introduce the set C consisting of continuous functions on the one-point compactification G_c of G; it becomes a Banach space with the norm $\|f\| = \max_g |f(g)|$. For an $f \in C$ and for an infinitesimal transformation $Y \in \Lambda$ we define

$$Yf = \lim_{h \downarrow 0} \frac{R_{\eta_Y(h)}f - f}{h},$$

where
$$\eta_Y(t) = \exp tY,$$

and where the limit should exist in the topology of C, i.e. uniformly on G_c. Denote by C_1 the set $f \in C$ for which Yf is defined. Let C_k be the set of f's such that $Y_1 \cdot Y_2 \ldots Y_k f$ is defined for any $Y_1, Y_2 \ldots Y_k \in \Lambda(G)$ and make it into a Banach space by introducing the norm

$$\|f\|_k = \|f\| + \sum_i \|X_i f\| + \ldots + \sum_{i_1 \ldots i_k} \|X_{i_1} X_{i_2} \ldots X_{i_k} f_k\|,$$

where we have chosen a basis $X_1, X_2, \ldots X_k$ in $\Lambda(G)$. In the same way we introduce the similar spaces C'_k starting from left (instead of right) translations. Both these spaces are dense in C. As a chart at e we use the canonical coordinates $x_1, x_2, \ldots x_d$ associated with $X_1, X_2, \ldots X_d$, and we will also assume that the x_i have been extended to G_c with $x_i \in C_2$. In a neighborhood of e we introduce the function

$$\varphi(g) = \sum_1^d x_i^2(g)$$

and it is possible to extend φ to G_c such that $\varphi \in C_2$ and is bounded away from zero outside the neighborhoods of e.

Introduce the functionals

$$\begin{cases} D_i f = \lim_{h \downarrow 0} \dfrac{f(\eta x_i(h)) - f(e)}{h} \\ D_{ij} f = \lim_{h \downarrow 0} \dfrac{f(\eta x_i(h)\eta x_j(h)) - f(\eta x_i(h)) - f(\eta x_j(h)) + f(e)}{h^2} \end{cases}$$

whenever these limits exist; in C_2 they give the values $X_i f(e)$ and $X_i X_j f(e)$ respectively.

Our object of study will be the Borel probability measures P on G_c and semi-groups P_t of such probability measures, $P_t * P_s = P_{s+t}$; $s, t \geq 0$. The P_t's should be continuous with continuity defined via weak convergence of probability measures on G_c (note that this is not the same thing as weak convergence on G), $P_h \to \delta_e$ as $h \downarrow 0$.

The corresponding operators

$$T_t f(g) = \int_{G_c} f(gh) P_t(dh)$$

are defined for $f \in C$ and form a strongly continuous semi-group.

4.2. Homogeneous processes on Lie groups

Unless the infinitesimal generator M of the semi-group T_t is a bounded operator we run into the usual difficulties because of the fact that M is not defined everywhere in C. But now we have at our disposal more detailed knowledge of the group, we can talk of differentiable functions etc. This makes it possible to describe M in terms of differential and integral operators. The interested reader should compare the following with the usual derivation of the Khinchin-Lévy representation on the real line.

We shall first show that M can be expressed in a certain way in a subset of C. Further we must show that M given in this subset really determines T_t.

Finally we shall show that any M of this form generates a continuous homogeneous process.

Consider a homogeneous stochastic process (see section 2.2) taking values in G_c and with the probability distributions P_t and the probability operators T_t. Since the T_t's form a strongly continuous semi-group the infinitesimal operator N of T_t is defined in a dense subset $\mathcal{D}$ of C. However, T_t is also strongly continuous in C_2' (see Notes 4.2.$_2$) and it will be convenient to introduce N' through

$$N'f = \lim_{h \downarrow 0} \frac{T_h f - f}{h}$$

in some subset $\mathcal{D}'$ dense in C_2'.

Now we modify the function $\varphi(g)$ slightly. For any $\varepsilon > 0$ it is possible to find a $\varphi' \in \mathcal{D} \cap C_2'$ such that $\|\varphi - \varphi'\|_2' < \varepsilon$, $\varphi'(e) = D_i \varphi' = D_{ij} \varphi' = 0$ for $i \neq j$ and $D_{ii} \varphi' = 2$ and φ' positive on $G_c - e$.

Now we shall prove that the limit

$$Af = \lim_{h \downarrow 0} \frac{T_h f(e) - f(e)}{h}$$

exists for every $f(g) \in \mathcal{E} =$ the class of functions in C that are twice continuously differentiable. For a given positive δ we choose an $f_\delta \in \mathcal{D} \cap C_2$ such that $|f(g) - f_\delta(g)| \leqslant \delta \varphi'(g)$; this requires an approximation reasoning of familiar type. But then

$$\frac{1}{h} \int_{G_c} [f(g) - f(e)] P_h(dg) = \frac{1}{h} \int_{G_c} [f(g) - f_\delta(g)] P_h(dg) + \frac{1}{h} \int_{G_c} [f_\delta(g) - f_\delta(e)] P_h(dg)$$

since $f_\delta(e) = f(e)$. The limit of the second term exists since $f_\delta \in \mathcal{D} \cap C_2$. But

$$\frac{1}{h} \int_{G_c} [f(g) - f_\delta(g)] P_h(dg) \leqslant \frac{\delta}{h} \int_{G_c} \varphi'(g) P_h(dg)$$

and since φ' is in $\mathcal{D}$ the limit of $1/h \int_{G_c} \varphi'(g) P_h(dg)$ exists. As δ can be made arbitrarily small the stated result follows. We can apply the above to the functions $x_i(g)$ and $x_i(g) x_j(g)$ and we will denote the resulting values by $b_i = A x_i$ and $b_{ij} = A x_i x_j$.—The reason why the functional A is so important is, as we shall see, that the infinitesimal generator of $\{T_t\}$ can be expressed through A.

Introduce now the measures depending upon the parameter t

$$\eta_t(E) = \frac{1}{t} \int_E \varphi(g) P_t(dg).$$

They are uniformly bounded (this follows from the fact that $\lim\limits_{t\downarrow 0} (1/t) \int_{G_c}\varphi(g)$ $P_t(dg)$ exists) and for $f\in\mathcal{E}$ the integrals

$$\int_{G_c} f(g)\eta_t(dg) = \frac{1}{t}\int_{G_c} f(g)\varphi(g)P_t(dg)$$

converge when t tends to zero. Thus the η_t converge to some measure η on G_c as t tends to zero so that

$$\lim_{h\downarrow 0}\frac{1}{t}\int_{G_c} f(g)\varphi(g)P_t(dg) = \int_{G_c} f(g)\eta(dg)$$

for any $f\in C$.

Now we are ready to represent A explicitly. If $f\in\mathcal{E}$ and if we denote $c=f(e)$, $c_i=D_i f$ and $c_{ij}=D_{ij}f$ then the function

$$h(g) = \frac{1}{\varphi(g)}[f(g) - c - \sum c_i x_i - \sum c_{ij}x_i x_j]$$

is in C and vanishes at $g=e$. But then we rewrite the definition of A

$$Af = A\left(c+\sum c_i x_i + \sum c_{ij}x_i x_j\right) + A\varphi h = \sum c_i b_i + \sum c_{ij}b_{ij} + A\varphi h$$

$$= \sum c_i b_i + \sum c_{ij}b_{ij} + \int_{G_c-e}\frac{1}{\varphi(g)}[f(g) - c - \sum c_i x_i(g) - \sum c_{ij}x_i(g)x_j(g)]\eta(dg).$$

The integrals $\int_{G_c-e}\dfrac{x_i x_j}{\varphi}\,\eta(dg)$ exist since the integrands are bounded and we can drop these terms if we agree to change the definition of the b_{ij}'s. Hence we can write

$$Af = \sum a_i D_i f + \sum a_{ij}D_{ij}f + \int_{G_c-e}\frac{1}{\varphi(g)}[f(g) - f(e) - \sum D_i f(g)x_i(g)]\eta(dg),$$

where the coefficients a_{ij} can be chosen symmetric in the indices, $a_{ij}=a_{ji}$ (this is possible since $D_{ij}-D_{ji}$ can be expressed linearly in the D_ν's). The matrix $\{a_{ij}\}$ must also be non-negative definite. Indeed, the operators T_t are non-negative and hence for any non-negative $f\in\mathcal{E}$ with $f(e)=D_i f=0$ we must have

$$\sum a_{ij}D_{ij} + \int_{G_c-e}\frac{1}{\varphi}f(g)\eta(dg) = \lim_{h\downarrow 0}\frac{1}{h}T_h f \geqslant 0.$$

Now choose an f with this behavior and $=[\sum z_i x_i(g)]^2\geqslant 0$ in a neighborhood of e but so small that the contribution of the integral is negligible. But then the quadratic form $\Sigma a_{ij}z_iz_j$ must be non-negative to avoid a contradiction. — Now let A be the functional described above but now restricted to C_2. The a_i, a_{ij} and η are chosen so that $Nf(e)=Af$. If N is the restriction to C_2 of the infinitesimal generator of some non-negative semi-group T'_t, then $T_t=T'_t$ since they both lead to A on C_2. To verify the last step we use a lemma that appears in many slightly different versions in the study of parabolic differential equations: Let $f(t,g)$ be a continuous function on $[0,\infty)\times G_c$ such that for any (t,g_0) satisfying

$$t>0, f(t,g_0)=\min_g f(t,g)$$

the derivative $\partial f/\partial t$ exists and is non-negative. Then $\min f(t,g)\geqslant\min f(0,g)$ for all t. See Notes 4.2.$_1$. In the present context we want to prove that the knowledge of A on C_2 determines T_tf for all t. It is enough to consider the f's which are constant near ω and belong to $\mathcal{E}$. But then $T_tf\in\mathcal{E}$ since $f\in C'_2$ and T_t maps C'_2 into itself (see Notes 4.2.$_2$). Now we interpret $f(0,g)$ as $f(g)$ and $f(t,g)$ as $T_tf(g)$. The conditions of the lemma are satisfied

$$\begin{cases} \dfrac{\partial f(t,\omega)}{\partial t}=0; \ t>0 \\[2mm] \dfrac{\partial f(t,g)}{\partial t}=\lim_{h\downarrow 0}\dfrac{T_hf(t,g)-f(t,g)}{h}=A\,(L_{g^{-1}}f(t,g)); \ g\in G, t>0. \end{cases}$$

Now if there were two functions $f_1(t,g)$ and $f_2(t,g)$ with these properties their difference $f_3=f_1-f_2$ would satisfy the above conditions and further $f_3(0,g)=0$, $g\in G$, $f_3(t,\omega)=0$, $t>0$. But if g_0 is a point where $f_3(t,g)$ attains a minimum then the first order term in A vanishes and the sum with second order terms is non-negative as well as the term with the integral. But then we must have $f_3(t,g)\geqslant\min_g f_3(0,g)=0$. Together with f_3 the function $-f_3$ must be a solution so that $-f_3\geqslant 0$ which gives us the desired result. Hence we know the form of the infinitesimal generator N of the semi-group T_t on C_2 and that N on C_2 determines the semi-group uniquely. It remains to show that any N of the form described above generates a semi-group T_t of probability operators.

For this purpose we shall use the following

Lemma 4.2.1. Consider a sequence of semi-groups of probability operators $T_t^n, n=1,2,\ldots$ Suppose that their infinitesimal generators exist on C_2 and are denoted by $M^nf=A_nL_{g^{-1}}$ and that for any $f\in C_2$ we have $M^nf\to Mf$ in the metric of C. For the corresponding η-measures we assume that $\eta_n(\omega)\to\eta(\omega)$.

Then $T_t^n f$ converges as $n \to \infty$ for every $f \in C$ to some $T_t f$, where T_t is a semi-group of probability operators whose infinitesimal generator exists on C_2 and coincides with M. We do not give the proof here since it is a bit laborious although along standard lines; see Notes 4.2.₃.

Now let M have the form

$$M f(g) = \int_{G_c - e} [f(gh) - f(g)] \eta(dh),$$

where η is a finite measure on $G_c - e$. But M is a bounded transformation of C into C and we know from chapter 3 that $T_t = \exp tM$, $t \geq 0$, forms a semi-group of probability operators continuous in the uniform operator topology with M as its infinitesimal operator. For a probabilistic interpretation of this semi-group see section 2.3.

Now consider an M of the form

$$M f(g) = \int_{G_c - e} [f(gh) - f(g) - \sum X_i f(g) x_i(h)] \eta(dh) + \sum a_i X_i f(g)$$

still with η a finite measure. Since the $x_i(h)$ are bounded we can write

$$M f(g) = X f(g) + \int_{G_c - e} [f(gh) - f(g)] \eta(dh)$$

with $X \in \Lambda(G)$. But we can always approximate Xf for every $f \in C_2$ by integral operators

$$\int_{G_c - e} [f(gh) - f(g)] \lambda_n(dh) \to X f(g)$$

in the topology of C and where the λ_n's are finite measures. Introduce now

$$M^n f(g) = \int_{G_c - e} [f(gh) - f(g)] [\eta(dh) + \lambda_n(dh)].$$

As the lemma above is applicable we see that M is the infinitesimal generator of a semi-group of probability operators.

To arrive at the general case we introduce the measures $\eta_n(dg) = [1 - e^{-n\varphi(g)}] \eta(dg)$. Then $\eta_n(dg)/\varphi(g)$ is a finite measure on $G_c - e$ and

$$M^n f(g) = \sum a_i X_i f(g) + \int_{G_c} [f(gh) - f(g)] \mu_n(dh)$$
$$+ \int_{G_c - e} [f(gh) - f(g) - \sum X_i f(g) x_i(h)] \frac{\eta_n(dh)}{\varphi(h)},$$

where μ_n is a finite measure such that

$$\int_{G_c} [f(gh) - f(g)]\mu_n(dh) \to \sum a_{ij} X_i X_j f(g)$$

for every $f \in C_2$ in the topology of C. But we know that M^n is the infinitesimal operator of some semi-group of probability operators. The lemma then tells us that M, as the limit of M^n, is also the infinitesimal generator of some semi-group of probability operators.

This leads us to the following theorem on the form of the infinitesimal generator of a semi-group of probability operators.

Theorem 4.2.1. Let T_t, $t \geqslant 0$, be a semi-group of probability operators on a separable Lie group. T_t has an infinitesimal generator M defined on C_2 through the expression

$$Mf(g) = \sum a_i X_i f(g) + \sum a_{ij} X_i X_j f(g)$$

$$+ \int_{G_c - e} [f(gh) - f(g) - \sum X_i f(g) x_i(h)] \frac{\eta(dh)}{\varphi(h)}.$$

The a_i are real numbers, the a_{ij} form a real, symmetric and non-negative definite $d \times d$ matrix and η is a finite measure on $G_c - e$. M on C_2 determines T_t.—Conversely if M is defined on C_2 as above then there is one and only one semi-group of probability operators T_t with M as its infinitesimal generator in C_2.

In analogy with what happens on the real line we introduce the following (see Notes 4.2.$_4$).

Definition 4.2.1. A probability measure P on a Lie group is said to be a normal (Gaussian) distribution if there is a semi-group of probability operators T_t with the infinitesimal generator M on C_2

$$Mf(g) = \sum a_i X_i f(g) + \sum a_{ij} X_i X_j f(g)$$

and

$$T_1 f(g) = \int_G f(gh) P(dh).$$

We now return for a moment to the compact groups. Consider a sequence of compact groups

$$\Gamma_1, \Gamma_2, \Gamma_3, \dots$$

with homomorphisms h_n of Γ_{n+1} on Γ_n. Introduce the set G of sequences $g = (\gamma_1, \gamma_2, \gamma_3, \dots)$ such that $\gamma_n \in \Gamma_n$, $\gamma_n = h_n(\gamma_{n+1})$ and define multiplication of two sequences $g' = (\gamma'_n)$, $g'' = (\gamma''_n)$ by

$$g = g' g'' = (\gamma_1' \gamma_1'', \; \gamma_2' \gamma_2'', \ldots).$$

Finally a topology is introduced into G through neighborhoods N, where N restricts a finite number of γ_n to fall into neighborhoods $N_n \subset \Gamma_n$. Then G is a compact group and is called the *limit of the groups* Γ_n.

Suppose now that we have defined continuous, homogeneous processes $P_t^{(n)}$ on Γ_n, $n = 1, 2, \ldots$ We demand a consistent definition so that $P_t^{(n)}$ is the probability distribution in Γ_n induced by $P_t^{(n+1)}$ through the homomorphism h_{n+1}. We then claim that this defines a continuous, homogeneous process in G.

First of all the marginal distributions $P_t^{(n)}$ determine a probability distribution P_t in G for every $t \geq 0$. Indeed let $f(g)$ be a continuous function on G depending only upon $\gamma_1, \ldots \gamma_n$. Then the value

$$Lf = \int_G f(\gamma_1 \ldots \gamma_n) \, P_t^{(n)}(d\gamma_1 \ldots \gamma_n)$$

is well defined. Using the compactness of G and uniform approximation by such functions L is directly extended to a positive functional on $C(G)$. This defines a regular probability measure P_t on G. Homogeneity, $P_{t+s} = P_t \ast P_s$, is shown in the same way. Finally continuity is seen to hold since if N is an arbitrary neighborhood of e involving $\gamma_1, \gamma_2, \ldots \gamma_n$, then

$$P_t(N) = P_t^{(n)}(N) \to 1 \text{ as } t \downarrow 0$$

and this is the same as continuity (see Notes 2.3.$_1$).

Let us note the following in passing. If we have for each n a sequence of probability measures $P_1^{(n)}, P_2^{(n)}, \ldots$ on Γ_n given consistently, they define a sequence $P_1, P_2, \ldots$ on G. Now if for every n the probability measures $P_1^{(n)}$, $P_2^{(n)}, \ldots$ converge to some $Q^{(n)}$, then this defines a $P \in \mathcal{P}(G)$ and $P_n \to P$. This can be proved by uniform approximation as above.

Now return to the homogeneous processes. A given compact group G can be represented as the limit of Lie groups Γ_n (see Notes 4.2.$_5$). But we know what form the homogeneous processes can take on Lie groups, i.e. we know how to characterize their infinitesimal generators, see Theorem 4.2.1. Given a continuous homogeneous process P_t on G the infinitesimal generator M_n of $P_t^{(n)}$ (the process induced in Γ_n by P_t) is of known form for every n. What is perhaps more interesting is that if we give an operator M that reduces to an operator M_n of the form mentioned for every n when we restrict ourselves to Γ_n, then the M_n's generate a set of processes $P_t^{(n)}$ consistently defined and we have seen above how this leads to a continuous, homogeneous process P_t on G.

It is probable that in concrete investigations of probability distributions on a compact group it will usually be convenient to base the study upon the unitary representations. Then it should be useful to see the connection with Lie groups.

4.3. The law of large numbers on stochastic Lie groups

Consider a triangular system of probability distributions on G_c

$$\begin{cases} P_{11} \\ P_{21},\ P_{22} \\ P_{31},\ P_{32},\ P_{33} \\ \quad \cdot \quad \cdot \quad \cdot \quad \cdot \quad \cdot \quad \cdot \end{cases}$$

One will assume in one form or another that the system is *infinitesimal*, i.e. that in any row the individual underlying stochastic variables are negligible, just as is done in the classical situation. More precisely we will let this mean that for any neighborhood N_0 of e we should have

$$\lim_{n \to \infty} \max_k\ P_{nk}(G_c - N_0) = 0.$$

Further it will be assumed that in any row $P_{ni} * P_{nj} = P_{nj} * P_{ni}$; in the classical cases such a commutativity assumption is, of course, automatically fullfilled. Note that it is also true trivially for the important case of *identical components*, i.e. when $P_{ni} = P_{nj}$ for all i and j.

Theorem 4.3.1. If the triangular system has these two properties and if it satisfies

1) $$\lim_{n \to \infty} \sum_{k=1}^{n} \left| \int_{N_0} x_i(g)\, P_{nk}(dg) \right| = 0,\ i = 1, 2, \ldots, d$$

2) $$\lim_{n \to \infty} \sum_{k=1}^{n} \int_{G_c} \varphi(g)\, P_{nk}(dg) = 0$$

then $P_n = P_{n1} * P_{n2} * \ldots * P_{nn}$ converges to the degenerate distribution having all its mass at the point e.

Proof: It will sufficient to prove that $T_n = T_{n1} T_{n2} \ldots T_{nn} \to I$ strongly. But as $\|T_n - I\| \leqslant 2$ it suffices to show that $\|(T_n - I)f\| \to 0$ for $f \in C_2$, since C_2 is dense in C.

Consider

$$[T_{nk} - I]f(g) = \int_{G_c - N_0} [f(gh) - f(g)]P_{nk}(dh) + \int_{N_0} [f(gh) - f(g)]P_{nk}(dh).$$

The first term is easily dealt with since

$$\left| \int_{G_c - N_0} [f(gh) - f(g)]P_{nk}(dh) \right| \leqslant 2\|f\|P_{nk}(G_c - N_0)$$

and condition 2) implies that

$$\lim_{n \to \infty} \sum_{k=1}^{n} P_{nk}(G_c - N_0) = 0.$$

To dominate the second term we expand the integrand in a Taylor series in N_0

$$f(gh) - f(g) = \sum_i X_i f(g) x_i(h) + \tfrac{1}{2}\sum_{ij} X_i X_j f(g\theta) x_i(h) x_j(h)$$

with easily understood notation. But then

$$\left| \int_{N_0} [f(gh) - f(g)]P_{nk}(dh) \right| \leqslant \sum_i |X_i f(g)| \cdot \left| \int_{N_0} x_i(h) P_{nk}(dh) \right|$$

$$+ \tfrac{1}{2}\sum_{i,j} \int_{N_0} |X_i X_j f(g\theta)| \cdot |x_i(h)| \cdot |x_j(h)| P_{nk}(dh)$$

$$\leqslant \|f\|_2 \sum \left| \int_{N_0} x_i(h) P_{nk}(dh) \right| + \text{const} \cdot \|f\|_2 \int_{N_0} \varphi(h) P_{nk}(dh).$$

Combining this with the conditions 1) and 2) we see that

$$\lim_{n \to \infty} \sum \|[T_{nk} - I]f\| = 0.$$

But

$$[T_n - I]f = [T_{n1} T_{n2} \ldots T_{nn} - I]f = [T_{n1} - I]f$$
$$+ [T_{n1} T_{n2} - T_{n1}]f + \ldots + [T_{n1} T_{n2} \ldots T_{nn} - T_{n1} T_{n2} \ldots T_{n\,n-1}]f$$

so that

$$\|[T_n - I]f\| \leqslant \sum_1^n \|T_{n1} T_{n2} \ldots T_{nk-1}(T_{nk} - I)f\| \leqslant \sum_1^n \|[T_{nk} - I]f\| \to 0$$

which completes the proof.

On the real line the law of large numbers is valid whatever the mean value is, as is obvious from the cummutativity of the real additive group. In the present context one would guess that replacing condition 1) by

$$\lim_{n\to\infty} \sum_{k=1}^{n} \int_{N_0} x_i(g)\, P_{nk}(dg) = a_i, \quad i = 1, 2, \ldots, d$$

the probability mass would converge to the element $\exp A$, where $A = \Sigma a_i X_i \in \Lambda(G)$. To prove this by direct extension of the previous proof does not seem possible since the lack of commutativity causes certain difficulties. We therefore have to postpone the discussion of this problem until we have probed a bit more deeply into this question.

4.4. A central limit theorem

If the factors in $T_n = T_{n1} T_{n2} \ldots T_{nn}$ or in $P_{n1} * P_{n2} * \ldots * P_{nn}$ are close to the identity one may hope that $T_{nk} \cong \exp(T_{nk} - I)$. Also we could then expect that the factors would commute approximately so that

$$V_n = \exp \sum_{k=1}^{n} (T_{nk} - I) \cong T_n.$$

A relation of this type is essential for the study of limit theorems on Lie groups. We shall prove the following

Theorem 4.4.1. The probability measures P_{nk} should form an infinitesimal and commutative triangular system such that

1) $$\sum_{k=1}^{n} \left| \int_{N_0} x_i(g)\, P_{nk}(dg) \right| \text{ is uniformly bounded,}$$

2) $$\sum_{k=1}^{n} \int_{G_c} \varphi(g)\, P_{nk}(dg) \text{ is uniformly bounded,}$$

3) For any positive ε there is a compact set $F = F_\varepsilon$ such that for all n we have

$$\sum_{k=1}^{n} P_{nk}(G - F) < \varepsilon.$$

Then $(T_n - V_n)f \to 0$ for any $f \in C$ in the C-norm.

Proof: With $S_{nk} = \exp(T_{nk} - I)$ we have, since for fixed n all the T_{nk} and S_{nk} commute,

$$(T_n - V_n)f = \sum_{k=1}^{n} T_{n1} \ldots T_{nk-1} S_{nk+1} \ldots S_{nn} (T_{nk} - S_{nk})f;$$

this is seen just as at the end of last section. Hence

$$\|(T_n - V_n)f\| \leqslant \sum_{k=1}^{n} \|(T_{nk} - S_{nk})f\|$$

since S_{nk} is a (Poisson) probability operator, $\|S_{nk}\| = 1$. But

$$T_{nk} - S_{nk} = T_{nk} - \left\{ I + (T_{nk} - I) + \frac{(T_{nk} - I)^2}{2!} + \ldots \right\} = -A_{nk}(T_{nk} - I)^2,$$

where
$$A_{nk} = \frac{1}{2!} I + \frac{(T_{nk} - I)}{3!} + \frac{(T_{nk} - I)^2}{4!} + \ldots,$$

so that the A_{nk}'s are operators of uniformly bounded norm. Hence

$$\|(T_n - V_n)f\| \leqslant \text{const.} \cdot \sum_{k=1}^{n} \|(T_{nk} - I)^2 f\|.$$

It is then enough to prove that the right side of the above inequality tends to zero as n tends to infinity for any $f \in C_2$, since C_2 is dense in C.

The rest of the proof is not difficult in principle although the amount of computational work will look overwhelming in a detailed proof. To give the reader the essence of the proof we shall assume that N is a neighborhood of e so small that the many Taylor expansions below are valid with the stated error terms. Let us write

$$(T_{nk} - I)^2 f(g)$$

$$= \int_{G_c} \left\{ \int_{G_c} [f(gxy) - f(gx) - f(gy) + f(g)] P_{nk}(dy) \right\} P_{nk}(dx)$$

$$= \int_{G_c - N} \int_{G_c - N} + \int_{G_c - N} \int_{N} + \int_{N} \int_{G_c - N} + \int_{N} \int_{N} = I_{1k} + I_{2k} + I_{3k} + I_{4k}.$$

To deal with I_{1k} we note that

$$|I_{1k}| \leqslant 4\|f\| P_{nk}^2 (G_c - N)$$

so that

$$\sum_{k=1}^{n} |I_{1k}| \leqslant 4\|f\| \sum_{k=1}^{n} P_{nk}(G_c - N) \cdot \max P_{nk}(G_c - N) \to 0$$

because of condition 2) and since the system $\{P_{nk}\}$ is infinitesimal.

To deal with I_{2k} we write

$$\int_{N} [f(gxy) - f(gx) - f(gy) + f(g)] P_{nk}(dy)$$

$$= \int_{N} [f(gxy) - f(gx)] P_{nk}(dy) - \int_{N} [f(gy) - f(g)] P_{nk}(dy)$$

and apply Taylor's theorem obtaining

$$\int_N [f(gh) - f(g)] P_{nk}(dh) = \sum_i X_i f(g) \int_N x_i(h) P_{nk}(dh)$$

$$+ \frac{1}{2} \int_N \sum_{i,j} X_i X_j f(g\theta) x_i(h) x_j(h) P_{nk}(dh)$$

so that

$$\left| \int_N [f(gh) - f(g)] P_{nk}(dh) \right| \leqslant \|f\|_1 \cdot \sum_i \left| \int_N x_i(h) P_{nk}(dh) \right|$$

$$+ \text{const} \cdot \|f\|_2 \int_N \varphi(h) P_{nk}(dh).$$

Hence, summing such terms,

$$\sum_{k=1}^n |I_{2k}| \leqslant 2\|f\|_2 \sum_{k=1}^n \left[\sum_i \left| \int_N x_i(h) P_{nk}(dh) \right| \right.$$

$$\left. + \text{const} \cdot \int_N \varphi(h) P_{nk}(dh) \right] P_{nk}(G_c - N)$$

which tends to zero as n tends to infinity.

In the fourth integral both integration variables x and y vary in the small neighborhood N and Taylor expansion and some patient calculation similar to the above shows that the contribution of this integral can be made small for large values of n. The third integral is more interesting

$$I_{3k} = \int_{G_c - N} \left\{ \int_N [f(gxy) - f(gy) - f(gx) + f(g)] P_{nk}(dx) \right\} P_{nk}(dy).$$

When we study the function

$$f(gxy) - f(gy) = f[gy(y^{-1}xy)] - f(gy),$$

where x varies in a small neighborhood of e, we cannot be sure that $y^{-1}xy$ varies in some small and fixed neighborhood of e for all $y \in G_c - N$. To take care of this difficulty condition 3) has been introduced. Let us write

$$I_{3k} = \left(\int_{G-F} + \int_{F-N} \right) \left\{ \int_N [f(gxy) - f(gx) - f(gy) + f(g)] P_{nk}(dx) \right\} P_{nk}(dy)$$

$$+ \int_N [f(g) - f(gx)] P_{nk}(dx) P_{nk}(\omega),$$

where the last term is the contribution from the element $y = \omega$. — The first integral gives no trouble since

$$\sum_{k=1}^{n}\left|\int_{G-F}\int_{N}\right|\leqslant 4\|f\|\sum_{k=1}^{n}P_{nk}(G-F)\leqslant 4\|f\|\varepsilon$$

according to the way the set F was introduced in the statement of the theorem.—The integrals

$$\int_{N}[f(g)-f(gx)]P_{nk}(dx)\,P_{nk}(\omega)$$

and

$$\int_{F-N}\left\{\int_{N}[f(gx)-f(g)]P_{nk}(dx)\right\}P_{nk}(dy)$$

are handled by standard series expansion again.—The remaining term is

$$\int_{F-N}\left\{\int_{N}[f(gxy)-f(gy)]P_{nk}(dx)\right\}P_{nk}(dy).$$

By series expansion

$$\int_{N}[f(gxy)-f(gy)]P_{nk}(x)\,|\leqslant\|f\|_{1}\sum_{i}\left|\int_{N}x_{i}(y^{-1}xy)\,P_{nk}(dx)\right|$$

$$+\,\text{const}\cdot\|f\|_{2}\int_{N}\varphi(y^{-1}xy)\,P_{nk}(dx).$$

Now the crucial point is that y varies only over a compact set. The maxima of the terms on the right side are then assumed for some values $y_{1},\dots y_{d}$ and y_{0}, so that

$$\sum_{k=1}^{n}\left|\int_{F-N}\int_{N}\right|\leqslant\|f\|_{2}\sum_{k=1}^{n}\left\{\sum_{i}\left|\int_{N}x_{i}(y_{i}^{-1}xy_{i})\,P_{nk}(dx)\right|\right.$$

$$\left.+\,\text{const}\cdot\int_{N}\varphi(y_{0}^{-1}xy_{0})\,P_{nk}(dx)\right\}P_{nk}(F-N)\to 0$$

as n tends to infinity since the system was infinitesimal. This completes the proof of the theorem.

We are now ready for a version of the central limit theorem.

Theorem 4.4.2. Let $\{P_{nk}\}$ be a commutative infinitesimal system of probability distributions on the Lie group such that

1) $\displaystyle\lim_{n\to\infty}\sum_{k=1}^{n}\int_{N_{0}}x_{i}(g)\,P_{nk}(dg)=a_{i},\quad \sum_{k=1}^{n}\left|\int_{N_{0}}x_{i}(g)\,P_{nk}(dg)\right|$ uniformly bounded,

2) $\displaystyle\lim_{n\to\infty}\sum_{k=1}^{n}\int_{N_{0}}x_{i}(g)\,x_{j}(g)\,P_{nk}(dg)=a_{ij}$

3)
$$\lim_{n\to\infty} \sum_{k=1}^{n} \int_{G_c - N_0} \varphi(g) P_{nk}(dg) = 0$$

for every neighborhood N_0 of e. Then $P_n = P_{n1} * P_{n2} * \ldots * P_{nn}$ converges to the normal distribution generated by the operator

$$M = \sum_i a_i X_i + \sum_{i,j} a_{ij} X_i X_j.$$

Proof: Using Theorem 4.4.1 we need only consider $P_n = \exp \sum_1^n (T_{nk} - I)$ and its limit when n tends to infinity. But P_n can be embedded in the semigroup $\exp t \sum_1^n (T_{nk} - I)$ of type compound Poisson. We can then write its infinitesimal generator as $(f \in C_2)$

$$\sum_{k=1}^{n} (T_{nk} - I) f(g) = \sum_{k=1}^{n} \int_{G_c} [f(gh) - f(g)] P_{nk}(dh)$$

$$= \sum_i a_i^{(n)} X_i f(g) + \int_{G_c - e} [f(gh) - f(g) - \sum_i X_i f(g) x_i(h)] \frac{\eta_n(dh)}{\varphi(h)}$$

with

$$\begin{cases} \eta_n(dh) = \sum_{k=1}^{n} \varphi(h) P_{nk}(dh). \\ a_i^{(n)} = \int_{N_0} x_i(h) \frac{\eta_n(dh)}{\varphi(h)} + \int_{G_c - N_0} x_i(h) \frac{\eta_n(dh)}{\varphi(h)}. \end{cases}$$

Now we use condition 1) to show that the first term in $a_i^{(n)}$ tends to a_i and condition 3) to show that the second term tends to zero. Further η_n tends to zero outside any neighborhood of e. Then it follows that

$$\sum_{k=1}^{n} (T_{nk} - I) f \to \sum_i a_i X_i f + \sum_{i,j} a_{ij} X_i X_j f$$

strongly, which implies (see Notes 4.2.3 and section 4.2) that P_n and T_n tend to $\exp \{\sum_i a_i X_i + \sum_{i,j} a_{ij} X_i X_j\}$ as stated.

We now return for a moment to the law of large numbers. Let us rephrase Theorem 4.3.1. as follows.

Theorem 4.3.1a. Let $\{P_{nk}\}$ be an infinitesimal and commutative system of probability distributions such that

1) $$\lim_{n\to\infty} \sum_{k=1}^{n} \int_{N_0} x_i(g) P_{nk}(dg) = a_i, \quad \sum_{k=1}^{n} \left| \int_{N_0} x_i(g) P_{nk}(dg) \right| \text{ uniformly bounded,}$$

2) $$\lim_{n\to\infty} \sum_{k=1}^{n} \int_{G_c} \varphi(g) P_{nk}(dg) = 0,$$

then $P_n = P_{n1} * P_{n2} * \ldots * P_{nn}$ converges to the degenerated distribution having all its mass at the point $\exp \sum_i a_i X_i$.

Proof: This reduces to the last theorem but with $a_{ij} \equiv 0$ so that P_n and T_n tend to $\exp \sum_i a_i X_i$ which is the probability operator corresponding to the probability distribution with its mass at the point $\exp \sum_i a_i X_i \in G$.

4.5. Illustration

4.5.1. The simplest Brownian motion is that on the real line, $x(t) = N(0, \sigma^2 t)$ if we assume no drift. Its frequency function

$$f(x, t) = \frac{1}{\sigma \sqrt{2\pi t}} \exp - \frac{x^2}{2\sigma^2 t}$$

satisfies the ordinary heat equation

$$\frac{\partial f}{\partial t} = \frac{\sigma^2}{2} \frac{\partial^2 f}{\partial x^2}$$

with the initial condition: all mass at $x = 0$ for $t = 0$. On the one-dimensional torus we have the same differential equation but we now identify x_1 and x_2 if $x_1 \equiv x_2 \pmod{2\pi}$. In other words we study the equation on the interval $(-\pi, \pi)$ with the boundary condition $f(-\pi, t) = f(\pi, t)$ which problem has the well-known solution

$$f(x, t) = \frac{1}{\sigma \sqrt{2\pi t}} \sum_{\nu = -\infty}^{\infty} \exp - \frac{(x - 2\nu\pi)^2}{2\sigma^2 t}.$$

This theta-series is simply the superposition of the previous solution at all points congruent to x modulo 2π. Note that R^1 is a covering group of T^1.

This simple fact generalizes considerably. Let G be a connected, locally simply connected group assumed locally compact and separable as usual. In the present context we may think of a separable, connected Lie group. Let G^* be a *covering group* of G with the homomorphism φ of G^* onto G. Denote by $N = \{g_1^*, g_2^*, \ldots\}$ the countable normal subgroup of G^* such that G is isomorphic to G^*/N, so that the image of $g \in G$ is $g_1^* g^*, g_2^* g^*, \ldots$ It is reasonable to choose G^* as the universal covering group of G; should we be interested in any other covering group Γ of G, then Γ is covered by G^*.

Now we start with a Brownian motion $g(t)$, $t \geqslant 0$, on G, see 3.1. We define a stochastic process on G^* sample function wise by following continuously, starting at the element e, the element in G^* corresponding to $g(t)$ for $0 \leqslant \tau \leqslant t$ and denote the endpoint of this curve by $g^*(t)$. Then $g^*(t)$ is also a Brownian motion with a probability measure P_t^* on G^*. It takes a certain amount of measure theoretic manipulations to make a complete proof of this statement but we leave this to the patient reader. The following is more interesting. Let S be a Borel set on G^*. Then $P_t^*(S)=0$ if and only if $P_t(\varphi S)=0$ since N is countable. Let G have the Haar measure dg and G^* have the Haar measure dg^*. Let P_t^* be absolutely continuous with respect to dg^*

$$P_t^*(E) = \int_E p^*(g^*, t)\, dg^*, \quad E \subset G^*.$$

For a given $g_0 \in G$ choose a neighborhood N^* of e^* so small that $g_n^* g_0^* N^*$ are disjoint for different n's (g_0^* is some inverse image of g_0) and N^* and $\varphi(N^*) = N$ are isomorphic. If E is any Borel set in $g_0 N$ then $\varphi^{-1}E$ is the union of the disjoint sets $g_n^*[\varphi^{-1}E \cap g_0^* N^*]$, so that

$$P_t(E) = \sum_1^\infty \int_{g_n^*[\varphi^{-1}E \cap g_0^* N^*]} p^*(t, g^*)\, dg^* = \sum_1^\infty \int_{\varphi^{-1}E \cap g_0^* N^*} p^*(t, g_n^* g^*)\, dg^*$$

$$= \int_{\varphi^{-1}E \cap g_0^* N^*} \sum_1^\infty p^*(t, g_n^* g^*)\, dg^*.$$

Since the sum in the integrand does not depend upon which image g^* we have chosen of g and since the Haar measure dg^* in $g_n^* N^*$ can be identified with the Haar measure dg in $g_0 N$ we get

$$P_t(E) = \int_E \sum_1^\infty p^*(t, g_n^* g^*)\, dg = \int_E p(t, g)\, dg$$

with
$$p(t, g) = \sum_1^\infty p^*(t, g_n^* g^*)$$

almost certainly. This is the desired generalization. We have used the left invariant Haar measures.

Let P be a normal distribution on T^1. Then P can be written as the convolution $P_1 * P_2$ where P_1 and P_2 are not normal distributions (see Notes 4.5.1.). We might say that Cramér's theorem does not hold on T^1. This has some interesting consequences. Think of the central limit theorem on the real line. We know that if it should hold each term in the sum must be either small or have approximately a normal distribution. On T^1 on the other hand we may very well have large and essentially non-normal terms and still arrive at the normal distribution.

4.5.2. Consider now the two-dimensional Lie group consisting of the 2×2 matrices

$$g = \begin{Bmatrix} 1 & g_1 \\ 0 & g_2 \end{Bmatrix}, \quad g_2 > 0,$$

the *affine group on the real line*. This simple but illuminating example will be studied from another point of view in 5.5.1, see also 7.5.1. To find the differential operator of the Brownian motion on this group it is convenient to derive the infinitesimal moments of the first and second order and write down the corresponding Fokker-Planck equation. We then get

$$g(t+h) = g(t) \begin{Bmatrix} 1 & \varepsilon_1(h) \\ 0 & 1 + \varepsilon_2(h) \end{Bmatrix},$$

where the ε's can be treated as normal stochastic variables for small $h > 0$. This is equivalent to

$$\left. \begin{aligned} g_1(t+h) &= \varepsilon_1(h) + g_1(t)[1 + \varepsilon_2(h)] \\ g_2(t+h) &= g_2(t)[1 + \varepsilon_2(h)] \end{aligned} \right\}$$

or

$$\left. \begin{aligned} g_1(t+h) - g_1(t) &= \varepsilon_1(h) + g_1(t)\varepsilon_2(h) \\ g_2(t+h) - g_2(t) &= g_2(t)\varepsilon_2(h) \end{aligned} \right\}.$$

The infinitesimal moments should then have the form

$$\left. \begin{aligned} &\lim_{h \downarrow 0} E \frac{\Delta g_1}{h} = a_1 + a_2 g_1 \\[2mm] &\lim_{h \downarrow 0} E \frac{\Delta g_2}{h} = a_2 g_2 \\[2mm] &\lim_{h \downarrow 0} E \frac{(\Delta g_1)^2}{h} = b_1 + b_2 g_1^2 + 2b_3 g_1 \\[2mm] &\lim_{h \downarrow 0} E \frac{(\Delta g_2)^2}{h} = b_2 g_2^2 \\[2mm] &\lim_{h \downarrow 0} E \frac{\Delta g_1 \Delta g_2}{h} = g_1 g_2 b_2 + g_2 b_3 \end{aligned} \right\}.$$

For convenience we specialize to the case $b_1 = b_2 = 1$ and $b_3 = 0$ which leads to the equation

$$\frac{\partial p}{\partial t} = \frac{1}{2} \frac{\partial^2}{\partial g_1^2} [(1 + g_1^2)p] + \frac{1}{2} \frac{\partial^2}{\partial g_2^2} [g_2^2 p] + \frac{\partial^2}{\partial g_1 \partial g_2} [g_1 g_2 p]$$

$$- \frac{\partial}{\partial g_1} [(a_1 + a_2 g_1)p] - \frac{\partial}{\partial g_2} [a_2 g_2 p].$$

To study the marginal distributions of g_1 and g_2 we just have to integrate out g_2 and g_1 respectively. The marginal frequency function p_2 of g_2 then satisfies

$$\frac{\partial p_2}{\partial t} = \frac{1}{2}\frac{\partial^2}{\partial g_2^2}(g_2^2 p_2) - \frac{\partial}{\partial g_2}(a_2 g_2 p_2).$$

This has the solution

$$p_2(g_2) = \frac{1}{\sqrt{2\pi t}\, g_2}\exp -\frac{(\log g_2 - (a_2 - \frac{1}{2}t)^2)}{2t}$$

which is not surprising since this is a lognormal frequency function as could be expected. The group operation from the g_2-coordinates point of view is simply multiplication and since we have to do with the limit of a great number of independent factors, each close to 1, we should expect to get a lognormal distribution. The marginal frequency function p_1 of g_1 satisfies

$$\frac{\partial p_1}{\partial t} = \frac{1}{2}\frac{\partial^2}{\partial g_1^2}[(1+g_1^2)p_2] - \frac{\partial}{\partial g_1}[(a_1 + a_2 g_1)p_2].$$

Now what is interesting is that there can be a stationary distribution for g_1 (but, of course, not for g_2, except trivially). Such a stationary frequency function would satisfy

$$\frac{1}{2}\frac{d^2}{dg_1^2}[(1+g_1^2)p_1] - \frac{d}{dg_1}[(a_1 + a_2 g_1)p_1] = 0$$

so that

$$\frac{1}{2}\frac{d}{dg_1}[(1+g_1^2)p_1] - (a_1 + a_2 g_1)p_1 = \tfrac{1}{2}(1+g_1^2)p_1' - [a_1 + (a_2 - 1)g_1]p_1 = C_1$$

with the general solution

$$p_1(g_1) = (1+g_1^2)^{a_2-1}\exp 2a_1 \,\mathrm{arctg}\, g_1 \cdot$$

$$\left[2C_1\int_0^{g_1}(1+h^2)^{-a_2}\exp(-2a_1\,\mathrm{arctg}\,h)\,du + C_2\right].$$

If $a_2 \geqslant \frac{1}{2}$ the above function is not integrable over $(-\infty, \infty)$ and hence not a frequency function. If $a_2 < \frac{1}{2}$ the integral inside the brackets behaves like $g_1^{-2a_2+1}$ so that to get a possible frequency function we must put $C_1 = 0$. Then we get the stationary frequency function

$$p_1(g_1) = C_2(1+g_1^2)^{a_2-1}\exp 2a_1\,\mathrm{arctg}\,g_1.$$

A consequence of this will be studied later.

4.5.3. Let us consider the group G of *linear fractional transformations*

$$z_1 \circ z = \frac{z_1 + z}{1 + \bar{z}_1 z}$$

leaving the open unit circle $C = \{z \mid |z| < 1\}$ invariant. A reader familiar with non-Euclidean geometry will recognize this as the parallell motions in the Lobachevsky plane represented by the circular model. Suppose now that in the group product $z_1 \circ z_2 \circ \ldots \circ z_n$ the factors are independent, all with the same probability distribution $P_{nk} = P_n$ and that the P_n's are rotationally invariant (isotropic) in C; then we can let $P_{nk} \to \delta_e$ and apply Theorem 4.4.2.

This leads to a Brownian motion on G corresponding to the Laplacian operator. In this special case the frequency function can be obtained explicitely. See Notes 4.5.3.

CHAPTER 5

LOCALLY COMPACT STOCHASTIC GROUPS

5.1. Unitary representations

For the two types of groups investigated in Chapter 3 we leaned heavily on the compactness or commutativity assumption and in Chapter 4 we used the differentiability throughout. Now, when we just assume the group to be locally compact, we can expect more difficulties to arise, but we also meet new problems that present a real challenge to the probabilists intuition. One of the main tasks of a general theory of stochastic groups should be to present a coherent picture of the probabilistic behavior of locally compact groups. At present our knowledge on this subject has serious gaps as will be painfully obvious to the reader when he reads the following chapter.

First let us remind the reader of some basic facts about representations, positive definite functions etc. of locally compact groups. Just as before we will consider separable groups only; this restriction may not be very essential and is made mainly for convenience. Let G be such a group and consider a *unitary representation* $r = (\mathcal{H}, U(g))$ where $\mathcal{H}$ is a separable Hilbert space and $U(g)$ is a unitary transformation of $\mathcal{H}$ for every $g \in G$. We demand that for any $z \in \mathcal{H}$ the vector valued function $U(g)z$ should be strongly continuous (note that this is equivalent to weakly continuous here) and $U(g)$ should satisfy the equation $U(g_1)U(g_2) = U(g_1 g_2)$ for all $g_1, g_2 \in G$. If there is no non-trivial, closed subspace of $\mathcal{H}$ left invariant by all $U(g)$, $g \in G$, then r is said to be *irreducible*. Just as for compact groups (see 3.1) we really study equivalence classes, where two different classes consist of non-equivalent representations. Note that for compact groups we can always choose the $\mathcal{H}$-space corresponding to an irreducible, unitary representation as finite dimensional unitary space and for commutative groups even as one-dimensional.

As an example let $\mathcal{H}$ be the Hilbert space of all functions $f(h)$ quadratically integrable with respect to left invariant Haar measure. Define $U(g)$ through $U(g)f(h) = L_g f(h) = f(g^{-1}h)$. Then $U(g)$ is a unitary (but in general not irreducible) representation and it is called the left *regular representation*.

Let $\{\mathcal{H}, U(g)\}$ be a unitary representation of G and introduce the complex valued function $\varphi(g) = (U(g)z, z)$ for some $z \in \mathcal{H}$. Then we have for any n and complex constants $c_1, c_2, \ldots, c_n$

$$\sum_{\nu,\mu=1}^{n} c_\nu \bar{c}_\mu \varphi(g_\mu^{-1} g_\nu) = \left\| \sum_{\nu=1}^{n} c_\nu U(g_\nu)z \right\|^2 \geqslant 0.$$

In analogy with the real line a function satisfying such inequalities is called *positive definite* on G. We will only consider continuous positive definite functions (as e.g. the one above). Inversely, any given continuous and positive definite function can be represented as $(U(g)z, z)$ for a suitable choice of $(\mathcal{H}, U(g))$ and z. This correspondence between positive definite functions and unitary representations will be used repeatedly in the following. If $\varphi(g) \not\equiv 0$ then $\varphi(e) > 0$ so that we can norm the positive definite functions by asking that $\varphi(e) = 1$ as will be done sometimes.

It is clear that the set Φ of positive definite functions is convex, $c\varphi_1 + (1-c)\varphi_2 \in \Phi$ for any $c \in (0,1)$ if $\varphi_1, \varphi_2 \in \Phi$. In Φ we will single out the extremal ones in the following way. Among the elements in Φ we use the partial ordering defined by $\varphi_1 \prec \varphi_2$ if $\varphi_2 - \varphi_1 \in \Phi$, and we say that φ_1 is subordinated to φ_2. A positive definite φ, for which the only subordinated functions are the multiples $c\varphi$, $c \in (0,1)$, is called an *elementary positive definite function*. It has been shown that they correspond exactly to the irreducible, unitary representations.

Some, but not all, positive definite functions can be constructed as follows by convolution. Let $a(g)$ be a function in $L(G)$, vanishing outside some compact set C, and form

$$p(g) = \int_G a(h)\bar{a}(hg)\nu(dh).$$

Then $p(g)$ is continuous and positive definite since

$$\sum_{i,j=1}^{n} c_i \bar{c}_j p(g_i g_j^{-1}) = \sum_{i,j=1}^{n} c_i \bar{c}_j \int_G a(h)\bar{a}(hg_i g_j^{-1})\nu(dh)$$

$$= \sum_{i,j=1}^{n} c_i \bar{c}_j \int_G a(kg_i^{-1})\bar{a}(kg_j^{-1})\nu(dk) = \int_G \left| \sum_{i=1}^{n} c_i a(kg_i^{-1}) \right|^2 \nu(dk) \geqslant 0.$$

Note also that $p(g)$ vanishes if $g \notin C^{-1}C$. The latter set is compact (see Notes 5.1.$_2$).

A fundamental result tells us that there exists a set $R = \{r\}$ of irreducible, unitary representations which is *complete* in the following sense: if g is an arbitrary element in G different from the unit element, then there is an $r \in R$ such that its $U(g_0) \neq I$. In other words, the irreducible, unitary representations are sufficiently many to be able to separate points in G.

This completeness property has important consequences. One can consider the elementary, positive definite functions as building blocks via their finite linear combinations $c_1\varphi_1 + c_2\varphi_2 + \ldots + c_n\varphi_n$, the so-called *trigonometric polynomials*. This fact will be useful to us and we formulate it in three different ways.

a) If μ is a bounded complex measure and

$$\int_{g \in G} \varphi(g)\mu(dg) = 0$$

for all elementary positive definite φ, then $\mu = 0$.

b) Any continuous function on G can be approximated uniformly on every compact set by trigonometric polynomials.

c) Any continuous positive definite function φ_0 can be approximated uniformly on G by function ψ of the form

$$\psi(g) = \int_{\varphi \in E} \varphi(g)\nu(d\varphi),$$

where E is a compact set of elementary, positive definite functions φ and ν is a Borel measure on E (see Notes 5.1.$_3$).

In the compact case it has been known for several decades that any unitary representation can be decomposed into a direct sum of irreducible, unitary representations. In the present case this discrete sum must be replaced by an integral. Let D be a set of elements δ and with a measure ϱ. To each $\delta \in D$ there should correspond a Hilbert space $\mathcal{H}_\delta$ with an inner product $(u,v)_\delta$. Now let us form the new Hilbert space denoted by

$$\mathcal{H} = \int_{\delta \in D} \mathcal{H}_\delta \sqrt{\varrho(d\delta)}$$

and consisting of functions $x = x_\delta$, $x_\delta \in \mathcal{H}_\delta$ for any $\delta \in D$, and of finite norm according to the inner product

$$(x, y) = \int_{\delta \in D} (x_\delta, y_\delta)_\delta \varrho(d\delta).$$

We call $\mathcal{H}$ the *direct integral* of the $\mathcal{H}_\delta$'s. It can be proved that any unitary representation is equivalent to a representation in a space $\mathcal{H}$ which is the direct product of $\mathcal{H}_\delta$'s; furthermore $U(g)x = \{U_\delta(g)x_\delta, \delta \in D\}$, where $U_\delta(g)$ is an irreducible, unitary transformation for almost every (ϱ) δ. The functions

$(U_\delta(g)x, y)$ depending upon g and δ are jointly measurable. This decomposition is not unique in general.

We shall use the following notion: G is said to be a P-group if the constant function 1 can be approximated uniformly on every compact subset by positive definite functions vanishing outside compact sets. If this approximation can be made by functions of the form $c \mathbin{*} \tilde{c}\,(g)$, where $\tilde{c}\,(g) = \overline{c(g^{-1})}$ and $c(g)$ vanishes outside a compact set then we speak of a P^1-group. More particularly one could try functions

$$p(g) = \frac{1}{\nu(C)}\, I_C \mathbin{*} \tilde{I}_C(g),$$

where $I_C(g)$ is the indicator function of a compact set C (see Notes 5.1.$_1$).

5.2. Fourier analysis of locally compact stochastic groups

In analogy to Chapter 3 we choose the following

Definition 5.2. On a locally compact group G the *Fourier transform* $\hat{P} = \hat{P}_r$ of a probability distribution $P \in \mathcal{P}(G)$ is defined by

$$\hat{P}_r z = \int_{g \in G} U(g) z P(dg),\ r \text{ irreducible} = (\mathcal{H}, U(g)),\ z \in \mathcal{H}.$$

Since $U(g)z$ is strongly continuous and bounded there is no difficulty in defining the integral; it can be understood as a Bochner integral. Note that the Fourier transform $\hat{P}_r$ is an operator valued function defined on the set of all unitary, irreducible representations of G.

One may note that the probability operator $Tf(g) = \int_G f(gh)P(dh)$ corresponds to the regular representation $f(g) \to R_h f(g) = f(gh)$, usually not irreducible.

We now have the following basic result.

Theorem 5.2.

a) For any irreducible, unitary representation r the Fourier transform $\hat{P}_r$ is a bounded linear operator in $\mathcal{H}$.

b) $\hat{P}_r = I$ if r is the identity representation.

c) P is uniquely determined if $\hat{P}_r$ is known for all irreducible, unitary representations r.

d) $\widehat{P_1 \mathbin{*} P_2} = \hat{P}_1 \cdot \hat{P}_2$.

e) Let $\bar{P}$ denote the probability distribution of g^{-1} if g has the distribution

P, so that $\bar{P}(E) = P(E^{-1})$. Then the Fourier transform of $\bar{P}$ is the adjoint $(\hat{P})^*$ of $\hat{P}$. In particular, the Fourier transform $\hat{P}$ is a self-adjoint operator if and only if P is a symmetric measure, $P = \bar{P}$. It is a normal operator if and only if $P * \bar{P} = \bar{P} * P$.

f) Given a sequence of probability distributions $P_1, P_2, \ldots$ converging weakly to $P \in \mathcal{P}(G)$: then $\hat{P}_n \to \hat{P}$ strongly. On the other hand if $\hat{P}_n \to \hat{P}$ (with $P \in \mathcal{P}(G)$) weakly then $P_n \to P$ weakly.

Proof: Statements a), b) and d) are obvious. To prove c) let P_1 and $P_2 \in \mathcal{P}(P)$ and $\hat{P}_1 = \hat{P}_2$ for all r. If $p(g)$ is an arbitrary, positive definite elementary function it can be written as $p(g) = (U(g)z, z)$ in terms of some irreducible $r = (\mathcal{H}, U(g))$. But then with $Q = P_1 - P_2$,

$$\int_G p(g) Q(dg) = (\hat{P}_1 z, z) - (\hat{P}_2 z, z) = 0$$

so that

$$\int_G \varphi(g) Q(dg) = 0$$

for any trigonometric polynomial $\varphi(g)$. This implies that (see section 5.1) $Q = 0$, $P_1 = P_2$.

To prove statement e) we consider

$$(\hat{\bar{P}}) = \int_G U(g) \bar{P}(dg) = \int_G U(g^{-1}) P(dg) = \int_G U^*(g) P(dg) = (\hat{P})^*.$$

The first part of statement f) is not difficult. For a given $x \in \mathcal{H}$ we have

$$\|(\hat{P}_n - \hat{P})x\|^2 = \|\hat{P}_n x\|^2 + \|\hat{P}x\| - 2Re(\hat{P}_n x, \hat{P}x).$$

Since the integrand is bounded and continuous we have

$$\lim_{n \to \infty} (\hat{P}_n x, \hat{P}x) = \lim_{n \to \infty} \int_G (U(g)x, \hat{P}x) P_n(dx) = \int_G (U(g)x, \hat{P}x) P(dx)$$

$$= (\hat{P}x, \hat{P}x) = \|\hat{P}x\|^2.$$

Since $P_n \to P$ weakly it follows that $\bar{P}_n \to \bar{P}$ and that $\bar{P}_n * P_n \to \bar{P} * P$ weakly. See Notes 2.2.$_2$. Hence

$$\lim_{n \to \infty} \|\hat{P}_n x\| = \lim_{n \to \infty} (\hat{P}_n x, \hat{P}_n x) = \lim_{n \to \infty} (\hat{P}_n^* \hat{P}_n x, x) = (\hat{P}^* \hat{P}x, x) = (\hat{P}x, \hat{P}x) = \|\hat{P}x\|^2$$

and this implies $(\hat{P}_n - \hat{P})x \to 0$ strongly.

The second part of statement f) deserves a careful discussion. Since

$P_n(G)=1=P(G)$ we have established weak convergence of P_n to P once we have shown vague convergence, i.e.

$$\lim_{n\to\infty} \int_G f(g) P_n(dg) = \int_G f(g) P(dg)$$

for every $f\in L(G)$. Let us first take $p(g)$ as an arbitrary (continuous) positive definite function. As usual we relate it to a unitary representation $\{\mathcal{H}, U(g)\}$ by $p(g)=(U(g)x, x)$. Decomposing $U(g)$ into irreducible components $U_\delta(g)$ we have (see section 5.1.)

$$p(g) = \int_{\delta\in D} (U_\delta(g)x_\delta, x_\delta)\varrho(d\delta)$$

and by Fubini's theorem and bounded convergence

$$\int_G p(g) P_n(dg) = \int_{\delta\in D} (\hat{P}_{n,\delta}x_\delta, x_\delta)\varrho(d\delta) \to \int_{\delta\in D} (\hat{P}_\delta x_\delta, x_\delta)\varrho(d\delta) = \int_G p(g) P(dg).$$

Now it remains for us to prove that this holds when p is replaced by an arbitrary function in $L(G)$. Let $f\in L$. Then f can be approximated uniformly on G by convolutions of the form $f*\tilde{\varepsilon}$ with $\varepsilon\in L(G)$, see Notes 5.2.$_4$. But $f*\tilde{\varepsilon}$ can be expressed linearly in terms of $(f\pm\varepsilon)*(\widetilde{f\pm\varepsilon})$ and $(f\pm i\varepsilon)*(\widetilde{f\pm i\varepsilon})$. The latter functions of the form $h*\tilde{h}$ are positive definite and continuous and this completes the proof.

The idempotents are described similarly to the compact and commutative groups.

Theorem 5.2.1. P is an idempotent probability measure on G if and only if it is the normed Haar-measure on a compact subgroup.

Proof: Since $(\hat{P}_r)^2=\hat{P}_r$ we have for any $x\in\mathcal{H}_r$, if we put $z=\hat{P}_r x$, that $\hat{P}_r z=z$. But just as before this implies that $U_r(g)z=z$ for any $g\in s(P)$. Since x was arbitrary this implies $U_r(g)\hat{P}_r=\hat{P}_r$ in $s(P)$ so that

$$s(P) \subset \bigcap_r \{g\,|\,U_r(g)\hat{P}_r = \hat{P}_r\}.$$

But the set on the right side is a closed subgroup of G. As usual we can assume that G has already been chosen as the smallest closed subgroup (of the original group) containing $s(P)$; then the right side above is simply G. Now consider the probability measure Q defined by $Q(E)=P(hE)$. Then

$$\hat{Q}_r = \int_G U_r(g)\,Q(dg) = \int_G U_r(g)\,P(hdg) = \int_G U_r(h^{-1}g)\,P(dg) = U_r(h^{-1})\hat{P}_r = \hat{P}_r.$$

Because of the uniqueness theorem it follows that $Q(E)=P(hE)=P(E)$: P is normed Haar-measure on G which is known to imply that G is compact. The rest of the proof is standard.

We have the following simple lemma.

Lemma 5.2.1. Let $P \in \mathcal{P}(G)$. If there is an irreducible representation $r = (\mathcal{H}, U(g))$ (different from the identity representation) such that $\hat{P}_r z = e^{i\theta} z$ for some nontrivial $z \in \mathcal{H}$, then the support $s(P)$ is contained in a coset of a closed, proper subgroup of G.

The proof is just a repetition of that used for Lemma 3.2.1. It should be mentioned though that the Lemma above is not stated in terms of the norm of the operator $\hat{P}_r$, but concerns possible eigen-values (of $\hat{P}_r$) of modulus 1. If we only assume that $\|\hat{P}\|=1$ then we know that there is a sequence of vectors $z_1, z_2, \dots \in \mathcal{H}, \|z_n\|=1$, such that $\|\hat{P}z_n\| \to 1$. Denoting $p_n(g) = (U(g)z_n, z_n)$ we have

$$\|\hat{P}z_n\|^2 = \left(\int_G U(g)z_n\,P(dg), \int_G U(g)z_n\,P(dg) \right)$$

$$= \int_G \int_G (U(g^{-1}h)z_n, z_n)\,P(dg)\,P(dh) = \int_G (U(g)z_n, z_n)\,Q(dg) = \int_G p_n(g)\,Q(dg),$$

where $Q = \bar{P} * P$. Since $p_n(e)=1$ and $|p_n(g)| \leqslant 1$ it follows that we can extract a subsequence n_ν such that

$$p_{n_\nu}(g) = (U(g)z_{n_\nu}, z_{n_\nu}) \to 1$$

almost everywhere with respect to Q-measure, or equivalently, $U(g)z_{n_\nu} - z_{n_\nu} \to 0$ strongly almost everywhere. But if this convergence holds for g_1 and g_2

$$\left. \begin{array}{l} U(g_1)z_{n_\nu} - z_{n_\nu} \to 0 \\[4pt] U(g_2)z_{n_\nu} - z_{n_\nu} \to 0 \end{array} \right\}$$

then $\qquad U(g_2^{-1})[U(g_1) - U(g_2)]z_{n_\nu} = U(g_2^{-1}g_1)z_{n_\nu} - z_{n_\nu} \to 0$

so that Q has all its mass contained in the subgroup defined by

$$\{g \mid U(g)z_{n_\nu} - z_{n_\nu} \to 0\}.$$

Unfortunately this is not very informative.

Parenthetically we mention that among the homogeneous processes on G we can characterize a special class of some limited interest as follows.

Theorem 5.2.2. Let P_t be a homogeneous process of discontinuous type on a locally compact group G. Then it is a compound Poisson process on G.

Proof and notation as in 2.3.

Before turning our attention to proper limit theorems we will make some remarks about convergence in probability for stochastic (locally compact) groups. Let $\gamma_n, n=1,2,\ldots$ be a sequence of stochastic elements taking values in G. Suppose that the sequence converges in probability to some stochastic element γ, i.e. for every neighborhood N of e and any $\varepsilon>0$ we have

$$P\{\gamma_n^{-1}\gamma \in N\} > 1-\varepsilon$$

for all n larger than some $n_0=n_0(N,\varepsilon)$. Then, of course, the double sequence $\gamma_n^{-1}\gamma_m$ converges in probability to e. What is more valuable is that the converse holds, a *Cauchy criterion*.

Theorem 5.2.3. If $\gamma_n^{-1}\gamma_m$ converges in probability to e as n and m tend to infinity, then the sequence γ_n is convergent in probability.

Proof: We can proceed as on the real line with the necessary modifications. Let us choose a positive sequence $\varepsilon_1, \varepsilon_2,\ldots$ with $\sum_1^\infty \varepsilon_\nu < \infty$, and a decreasing sequence $N_1, N_2,\ldots$ of neighborhoods of e with $N_\nu^2 \subset N_{\nu-1}$ and $\underset{\nu}{\cap} N_\nu = e$. Let us choose the integers n_ν such that

$$P\{\gamma_{n_\nu}^{-1}\gamma_{n_{\nu+1}} \in N_\nu\} > 1-\varepsilon_\nu.$$

Consider the sets

$$S_r = \overset{\infty}{\underset{\nu=r}{\cap}} \{\gamma_{n_\nu}^{-1}\gamma_{n_{\nu+1}} \in N_\nu\}$$

and

$$S = \overset{\infty}{\underset{r=1}{\cup}} S_r.$$

Then

$$P(S_r^*) \leqslant \overset{\infty}{\underset{\nu=r}{\sum}} P\{\gamma_{n_\nu}^{-1}\gamma_{n_{\nu+1}} \notin N_\nu\} \leqslant \overset{\infty}{\underset{\nu=r}{\sum}} \varepsilon_\nu$$

so that

$$P(S_r) \geqslant 1 - \overset{\infty}{\underset{\nu=r}{\sum}} \varepsilon_\nu \uparrow 1$$

as r tends to infinity, so that $P(S)=1$. But if the event S occurs, which is almost certain, then γ_{n_ν} is a Cauchy sequence. Indeed, if N and $\varepsilon>0$ are given then we can find an integer p such that $N_p \subset N$ and (for $\nu<\mu$), using the relation $N_\nu^2 \subset N_{\nu-1}$, we get

$$\gamma_{n_\nu}^{-1}\gamma_{n_\mu} = \gamma_{n_\nu}^{-1}\gamma_{n_{\nu+1}}\gamma_{n_{\nu+1}}^{-1}\gamma_{n_{\nu+2}}\cdots\gamma_{n_{\mu-1}}^{-1}\gamma_{n_\mu} \in N_\nu N_{\nu+1}\ldots N_{\mu-1} \subset N_{\nu-1} \subset N_p \subset N$$

for all $\nu, \mu > p$. Every locally compact topological group is complete so that there is an element γ (see Notes 5.2.$_1$) such that $\gamma_{n_\nu} \to \gamma$. But since γ_{n_ν} converges in probability to γ as ν tends to infinity and since $\gamma_n^{-1}\gamma = \gamma_n^{-1}\gamma_{n_\nu}\gamma_{n_\nu}^{-1}\gamma$ it follows from $\gamma_n^{-1}\gamma_{n_\nu} \to e$ that γ_n converges in probability to γ.

Note that we have also obtained as a by-product the existence of a subsequence converging almost certainly.

Now let $g_1, g_2, \ldots$ be independent stochastic group elements on G. We do not assume that they all have the same distribution. Form the product $\gamma_n = g_1 g_2 \ldots g_n$. It would be convenient to have an analytical criterion that this product would converge.

Theorem 5.2.4. The products $\gamma_n = g_1 g_2 \ldots g_n$ converge in probability as n tends to infinity if for all of the irreducible representations

$$\sum_{n=1}^{\infty} \|\hat{P}_n - I\| < \infty,$$

where $\hat{P}_n$ is the Fourier transform of g_n.

Proof: Writing $\hat{P}_n = I + \Delta_n$ we consider the Fourier transform $Q_{n,m}$ of the stochastic group element $g_{n+1} \cdot g_{n+2} \cdots g_m$

$$Q_{n,m} = \hat{P}_{n+1} \cdot \hat{P}_{n+2} \ldots \hat{P}_m = \prod_{n+1}^{m} (I + \Delta_\nu) = I + \sum_{\nu=n+1}^{m} \Delta_\nu + \sum_{n+1 \leqslant \nu < \mu \leqslant m} \Delta_\nu \Delta_\mu + \ldots$$

Thus we have with $d_n = \|\Delta_n\|$

$$\|Q_{n,m} - I\| \leqslant \sum_{n+1}^{m} d_\nu + \tfrac{1}{2}\left(\sum_{n+1}^{m} d_\nu\right)^2 + \tfrac{1}{6}\left(\sum_{n+1}^{m} d_\nu\right)^3 + \ldots = \exp\left(\sum_{n+1}^{m} d_\nu\right) - 1.$$

As n tends to infinity $\sum_{n+1}^{m} d_\nu$ tends to zero (by the assumption of the theorem) and $Q_{n,m}$ tends to I, the Fourier transform of δ_e. As $\gamma_n^{-1}\gamma_m = g_{n+1}g_{n+2}\cdots g_m$ tends to e in probability as n and m tend to infinity we can appeal to Theorem 5.2.3. and the statement of the Theorem 5.2.4. follows (see Notes 5.2.$_2$).

One more remark about products of independent group elements. We have a number of modes of convergence: weak convergence of probability distributions, convergence in probability, convergence almost certainly etc. On the real line one knows fairly well the logical relations between the different types of convergence and much of this carries over to general locally compact and separable groups with minor modifications only. It should be noted that this is not always so, and we make the following observation. It is known (see Notes 5.2.$_3$) that for a sum $\sum_1^{\infty} x_n$ of independent real-valued stochastic variables x_n convergence almost certainly is equivalent to convergence of the distributions of the partial sums. Now let G be a compact group, let g_1 be a stochastic element distributed according to the normed Haar measure, and let $g_2, g_3, \ldots$ be a sequence of independent stochastic group elements such that $\prod_2^\infty g_n$ does not converge almost certainly. Then the distribution of the partial products $\prod_1^n g_\nu$ is just the normed Haar measure, while the partial products themselves do not converge almost certainly.

5.3. Limit theorems on locally compact stochastic groups

In this section we shall obtain some limit theorems for convolutions on locally compact groups (see also the beginning of Chapter 3). We believe that the methods used can be applied in much more general situations and that they should give a great variety of relevant limit theorems.

For a given element $g \in G$ the unitary operator $U(g)$ corresponding to some irreducible, unitary representation $(\mathcal{H}, U(g))$ can be written as (see Notes 5.3.$_1$).

$$U(g) = \int_{-\pi}^{\pi} e^{i\lambda} d\, E_g(\lambda),$$

where $E_g(\lambda)$ is a resolution of the identity (see Notes). This resolution is essentially unique when $U(g)$ is given. Introducing the self-adjoint operator

$$H(g) = \int_{-\pi}^{\pi} \lambda d\, E_g(\lambda),$$

of bounded norm, $\|H(g)\| \leqslant \pi$, we can represent $U(g)$ as $U(g) = \exp i H(g)$. Note that we cannot claim that $H(g_1 g_2) = H(g_1) + H(g_2)$ because of the lack of commutativity. But we know from perturbation theory that, since $U(g)$ is continuous, $E_g(\lambda)z$ depends continuously upon g except possibly at the jumps of $E_g(\lambda)$. This implies that $H(g)$ is strongly continuous over G. Alternatively we could introduce the family of operators

$$N(\lambda) = \int_G E_g(\lambda) P(dg).$$

Note that $N(\lambda)$ is no longer necessarily a resolution of the identity. Anyway it is non-decreasing in λ with the usual partial ordering of linear operators in a Hilbert space (see Notes 5.3.$_3$). Then the operator $H(P)$ below can be written as

$$H(P) = \int_{-\pi}^{\pi} \lambda N(d\lambda).$$

For any probability distribution P we can form the self-adjoint operator $H(P)$ defined through

$$H(P)z = \int_G H(g)z P(dg),$$

say, with the right member defined as a Bochner integral (see Notes $5.3._2$), and its norm is at most π. This operator plays a similar rôle to that of the ordinary mean value on the real line (or in R^k or some other linear vector space) but the similarity does not extend very far since we do not have scalar multiplication nor the commutativity property. Anyway it does not seem unreasonable *a priori* to introduce the concept of a mean value through the following tentative definition that will be modified later on. Given a probability measure $P \in \mathcal{P}(G)$ on a locally compact group G we say that the element $\bar{g} \in G$ is the mean value of P on G if

$$H(\bar{g}) = H(P) = \int_G H(g) P(dg)$$

for every irreducible, unitary representation.

The uniqueness of this definition is clear, since if $H(\bar{g}_1) = H(P) = H(\bar{g}_2)$, then $U(\bar{g}_1) = U(\bar{g}_2)$ for every $(\mathcal{H}, U(g))$ which is known to imply $\bar{g}_1 = \bar{g}_2$.

The existence of a mean value is not ensured. It should be noted here that we have expressed U as $\exp iH$ with a bounded H. There are situations however, where it is more natural to work with representations with unbounded H. Then of course the integral defining $H(P)$ needs a more careful study and some sort of integrability condition on P is needed. We will return to this question in the next section. The reader will find it illuminating to compare with the simple case $G = R^k$.

If P is symmetric then $H(P) = 0$, since $H(g^{-1}) = -H(g)$, and $\bar{g} = e$.

This concept of mean value is actually of no practical interest, but it can be modified, and we will phrase Theorem 5.3.1. in different although related terms. Let us consider a triangular array of stochastic group elements

$$
\begin{cases}
g_1 \\
g_{21}, \ g_{22} \\
g_{31}, \ g_{32}, \ g_{33} \\
\quad \cdot \quad \cdot \quad \cdot \quad \cdot \quad \cdot \quad \cdot \\
g_{n1}, \ g_{n2}, \ldots, \ g_{nn} \\
\quad \cdot \quad \cdot \quad \cdot \quad \cdot \quad \cdot \quad \cdot,
\end{cases}
$$

where the stochastic elements are independent and all of them in the nth row have the same distribution P_n. We have already considered the operator

$$N_1 = H(P) = \int_{-\pi}^{\pi} \lambda N(d\lambda).$$

Introduce also
$$N_2 = \int_{-\pi}^{\pi} \lambda^2 N(d\lambda).$$

Theorem 5.3.1. Assume that

1)
$$H(P_n) = \int_G H(g) P_n(dg) = \frac{H(g_0)}{n} + R_n, \quad \|R_n\| = o\left(\frac{1}{n}\right),$$

and

2)
$$\|N_2^{(n)}\| = \left\| \int_G H^2(g) P_n(dg) \right\| = o\left(\frac{1}{n}\right).$$

Then
$$\lim_{n\to\infty} P_n^{n*} = \delta_{g_0}.$$

Proof: The Fourier transform of P_n^{n*} is

$$(\hat{P}_n)^n = \left(\int_G U(g) P_n(dg) \right)^n = \left(\int_G \exp\, iH(g) P_n(dg) \right)^n = (I + iH(P_n) + A_n)^n,$$

with
$$A_n = \int_{-\pi}^{\pi} [e^{i\lambda} - 1 - i\lambda] N(d\lambda) = \int_{-\pi}^{\pi} [\cos\lambda - 1] N(d\lambda)$$
$$+ i \int_{-\pi}^{\pi} [\sin\lambda - \lambda] N(d\lambda) = A_n' + A_n''.$$

Since $|\cos\lambda - 1|$ and $|\sin\lambda - \lambda| \leqslant a\lambda^2$ and since the operators $N(d\lambda)$ are non-negative we have

$$\left.\begin{array}{l} -aN_2^{(n)} \leqslant A_n' \leqslant aN_2^{(n)} \\ -aN_2^{(n)} \leqslant A_n'' \leqslant aN_2^{(n)} \end{array}\right\}.$$

But this implies that

$$\|A_n'\|, \|A_n''\| \leqslant a\|N_2^{(n)}\| = o\left(\frac{1}{n}\right).$$

Hence $\|A_n\| = o\,(1/n)$ and

$$(\hat{P}_n)^n = \left[I + \frac{iH(g_0)}{n} + o\left(\frac{1}{n}\right) \right]^n,$$

where $o\,(1/n)$ stands for an operator of norm $o\,(1/n)$. As n tends to infinity

$$(\hat{P}_n)^n \to \exp\, iH(g_0) = U(g_0) = \hat{\delta}_{g_0}$$

which concludes the proof.

Now we also consider the operator

$$N_3^{(n)} = \int_{-\pi}^{\pi} |\lambda|^3 N\,(d\lambda).$$

The reader will observe that these operators N_1, N_2 etc. play the same rôle here as do the ordinary moments in the classical theory.

Theorem 5.3.2. Assume that the stochastic elements have mean value e and probability distributions satisfying the conditions

1) $$N_2^{(n)} = \int_{-\pi}^{\pi} \lambda^2 N\,(d\lambda) = \frac{S}{n} + o\left(\frac{1}{n}\right)$$

2) $$N_3^{(n)} = \int_{-\pi}^{\pi} |\lambda|^3 N\,(d\lambda) = o\left(\frac{1}{n}\right).$$

If $\pi \in \mathcal{D}(G)$ is a probability distribution with $\hat{\pi} = \exp\,(-S/2)$ then $P_n^{n*} \to \pi$.

Proof: We have in the same way as in the previous proof

$$\hat{P}_n = \int_{-\pi}^{\pi} e^{i\lambda} N\,(d\lambda)$$

$$= \int_{-\pi}^{\pi} \left[1 + i\lambda - \frac{\lambda^2}{2}\right] N\,(d\lambda) + \int_{-\pi}^{\pi} \left[e^{i\lambda} - 1 - i\lambda + \frac{\lambda^2}{2}\right] N\,(d\lambda) = I - \frac{S}{2n} + o\left(\frac{1}{n}\right)$$

so that $$(\hat{P}_n)^n \to \exp\left(-\frac{S}{2}\right) = \hat{\pi}.$$

Note that since S must be a non-negative operator there is a self-adjoint operator B so that $S = B^2$. Then $\hat{\pi}$ takes the familiar-looking form exp $(-B^2/2)$.

In the classical limit theorems the distributions $P_{n\nu}$ are often defined through normed stochastic variables $b_{n\nu}(x_{n\nu} - a_{n\nu})$. In the present very general situation *we do not have access to such norming operations* as multiplication by a scalar $b_{n\nu}$. This is a basic difficulty in developing this theory and one that has to be resolved before any definitive general theory can be established. Given a sequence of stochastic variables $x_1, x_2, x_3, \ldots$ (for concreteness we can think of $x_n = g_1 g_2 \ldots g_n$ on a stochastic group) defined in some space X. To norm the distributions of the x's we map X upon some other space Y via a sequence of functions $y_n = y_n(x)$, and a limit theorem will state that the distribution of $y_n(x)$ converges as n tends to infinity. But such normings can be made in many various ways that can be of different value to us. Suppose that we can prove two different limit theorems $\mathcal{L}^1$ and $\mathcal{L}^2$ via the two mappings $y_n^1 = y_n^1(x)$ and $y_n^2 = y_n^2(x)$

$$\begin{cases} \mathcal{L}^1 : \lim_{n\to\infty} \text{distribution of } y_n^1(x) = D^1, \\ \mathcal{L}^2 : \lim_{n\to\infty} \text{distribution of } y_n^2(x) = D^2. \end{cases}$$

If D^1 is non-atomic and D^2 has at least one atom, say A, then we will probably consider the limit theorem $\mathcal{L}^1$ as superior to $\mathcal{L}^2$. Indeed, the limit theorem $\mathcal{L}^2$ lumps some probability in A, while this probability is resolved by $\mathcal{L}^1$, so that $\mathcal{L}^1$ can be said to be *more detailed* than $\mathcal{L}^2$.

Another possibility would be to consider one limit theorem $\mathcal{L}^1$ as *at least as good as* another $\mathcal{L}^2$ if D^1 is absolutely continuous with respect to D^2. This would define a partial ordering among all the $\mathcal{L}$'s.

The practical interest in a limit theorem is that we will get approximations of the type

$$P_n(x_n \in S_n) = P_n(y_n \in S) \cong D(S),$$

where $S_n = y_n^{-1}(S)$ and the mapping is assumed to be one-one.

In the next section we shall consider another sort of norming that far from solves the above problem generally but still leads to limit theorems of interest.

5.4. Limit theorems on certain divisible groups

On the real line (and in R^k and in a Banach space) the normed sums

$$\frac{1}{n}x_1 + \frac{1}{n}x_2 + \dots + \frac{1}{n}x_n \text{ and } \frac{1}{\sqrt{n}}y_1 + \frac{1}{\sqrt{n}}y_2 + \dots + \frac{1}{\sqrt{n}}y_n, \, y_\nu = x_\nu - Ex_\nu,$$

have particularly attractive limit properties. On a general locally compact group we do not have any operation corresponding to scalar multiplication $c \cdot x$ in a linear vector space. This causes an essential difficulty which will be discussed briefly later on in this section.

Let us consider groups for which the nth roots are uniquely determined, i.e. for every $g \in G$ and natural number n there is a unique element $h \in G$ satisfying the equation $h^n = g$; h is denoted by $g^{1/n}$. Such groups are sometimes referred to as divisible R-groups. This means a real restriction although it is true that one can sometimes embed a given group in one with this property. The rational powers g^r, $r = $ rational number, can then be defined, and if $g^r \to e$ as $r \to 0$ we can introduce arbitrary powers g^t, t real. We shall assume the following:

a) For any real t and $g \in G$ there is an element $g^t \in G$, depending continuously upon g and t.

b) $g^0 = e$, $g^1 = g$.

c) $g^{t+s} = g^t g^s$.

We shall now introduce the concept of *average*, different from the mean value of section 5.3. Let $\{\mathcal{H}, U(g)\}$ be an arbitrary, irreducible and unitary representation of G. For a fixed g the operators $V_t = U(g^t)$ form a continuous group of unitary operators

$$V_{t+s} = U(g^{t+s}) = U(g^t g^s) = U(g^t) U(g^s) = V_t V_s.$$

We then know that (see Notes 5.4.$_1$)

$$V_t = \exp\, itH(g) = \int_{-\infty}^{\infty} e^{it\lambda}\, dF_g(\lambda),$$

where $F_g(\lambda)$ is a resolution of the identity associated with the self-adjoint operator

$$H(g) = \int_{-\infty}^{\infty} \lambda\, dF_g(\lambda).$$

The operators $H(g)$ are perhaps not defined in the whole $\mathcal{H}$ since they need not be bounded. We shall assume that there is a set $\mathcal{D}$ everywhere dense in $\mathcal{H}$ such that $H(g)z$ is defined for $z \in \mathcal{D}$ and all $g \in G$ and further that

$$\int_G \|H(g)z\|\, P(dg) < \infty.$$

Then we can introduce the operator $H(P)$ in $\mathcal{D}$ by

$$H(P)z = \int_G H(g)z P(dg).$$

Definition 5.4.1. If there exists an element $\bar{g} \in G$ such that $H(\bar{g})z = H(P)z$ for all $z \in \mathcal{D}$, then $\bar{g}$ is called the average of P over G.

If $\bar{g}$ exists it is uniquely determined. Indeed, if $H(\bar{g}_1) = H(P) = H(\bar{g}_2)$, then the operators $e^{iH(\bar{g}_1)}$ and $e^{iH(\bar{g}_2)}$ are unitary ($= U(\bar{g}_1)$ and $U(\bar{g}_2)$) and defined everywhere and equal in $\mathcal{H}$, hence $\bar{g}_1 = \bar{g}_2$.

If P has all its mass in g_0, then P's average is just g_0.

The average of the stochastic group element g^t, $t = $ a real constant, is equal to $(\bar{g})^t$ if $\bar{g}$ exists.

Given two probability distributions P_1 and P_2 on G such that the operators $H(P_1)$ and $H(P_2)$ commute and such that their averages $\bar{g}_1$ and $\bar{g}_2$ exist, then the average of the distribution $p_1 P_1 + p_2 P_2$, $0 \leqslant p_1$, $p_2 \leqslant 1$, $p_1 + p_2 = 1$, is equal to $(\bar{g}_1)^{p_1}(\bar{g}_2)^{p_2}$. In particular if a distribution P has the average $\bar{g}$ then the "reduced" distribution $\frac{1}{2}P + \delta_{(\bar{g})^{-1}}$ has the average e.

As pointed out above the operators $H(g)$ need not be bounded and this introduces additional difficulties into the problem. In exceptional cases— such as when $M(\lambda)$ below has all its variation in a finite interval—we may avoid this, but otherwise we will have to be prepared to devote a good deal of time and effort to settle this question.

Let us state some simple but possibly useful results. The Fourier transform of the variable $g^{1/n}$ is based on the representations

$$U(g^{1/n}) = \int_{-\infty}^{\infty} e^{i(\lambda/n)} F_g(d\lambda) = \int_{-\pi}^{\pi} e^{i\mu} E_g(d\lambda)$$

with

$$E_g(\lambda'') - E_g(\lambda') = \sum_{k=-\infty}^{\infty} [F_g(n\lambda'' + 2\pi kn) - F_g(n\lambda' + 2\pi kn)]$$

for $-\pi \leqslant \lambda' < \lambda'' < \pi$. The operator $N_1 = H(P_n)$ associated with the stochastic element $g^{1/n}$ then takes the form

$$N_1 = H(P_n) = \int_{-\pi}^{\pi} \lambda N^{(n)}(d\lambda) = \int_{-\infty}^{\infty} \left[\frac{\lambda}{n}\right] M(d\lambda),$$

where

$$M(\lambda) = \int_G F_g(\lambda) P(dg).$$

Remember that P is the probability distribution of the stochastic group element g. By $[x]$ we denote the periodic function with period 2π and defined by $[x] = x$ in the interval $[-\pi, \pi)$. In the same way we get

$$N_2 = \int_{-\pi}^{\pi} \lambda^2 N^{(n)}(d\lambda) = \int_{-\infty}^{\infty} \left[\frac{\lambda}{n}\right]^2 M(d\lambda).$$

Recalling Theorem 5.3.1, we get immediately

Theorem 5.4.1. Assume that

1) $\qquad M_{[1]} = \int_{-\infty}^{\infty} \left[\frac{\lambda}{n}\right] M(d\lambda) = \frac{1}{n} \int_{-\infty}^{\infty} \left[\frac{\lambda}{n}\right] F_{g_0}(d\lambda) + R_n, \ \|R_n\| = o\left(\frac{1}{n}\right),$

and

2) $\qquad \|M_{[2]}\| = \left\| \int_{-\infty}^{\infty} \left[\frac{\lambda}{n}\right]^2 M(d\lambda) \right\| = o\left(\frac{1}{n}\right).$

Then $g_2^{1/n} g_2^{1/n} \ldots g_n^{1/n} \to g_0$ in probability, or equivalently, $P_n^{n*} \to \delta_{g_0}$.

As an illustration we state without proof a corollary to this theorem. Let P be a symmetric probability distribution with

a)
$$\left\| \int_{-\pi n}^{\pi n} \lambda^2 M(d\lambda) \right\| = o(n)$$

b)
$$\left\| \int_{\pi n}^{\infty} M(d\lambda) \right\| = o\left(\frac{1}{n}\right).$$

Then $P_n^{n*} \to \delta_e$.

We can also get a rather unsatisfactory extension of Theorem 5.3.2.

Theorem 5.4.2. Assume that there is a positive operator $\hat{\pi}$, such that

1)
$$\|M_{[1]}\| = o\left(\frac{1}{n}\right)$$

2)
$$M_{[2]} = I - (\hat{\pi})^{1/n} + o\left(\frac{1}{n}\right)$$

3)
$$\|M_{[3]}\| = \left\| \int_{-\infty}^{\infty} \left|\left[\frac{\lambda}{n}\right]\right|^3 M(d\lambda) \right\| = o\left(\frac{1}{n}\right).$$

Then $P_n^{n*} \to \pi$ weakly.

Proof: Compare with the proof of Theorems 5.3.1. and 5.3.2. and note that

$$(\hat{P}_n)^n = \left[I + M_{[2]} + o\left(\frac{1}{n}\right) \right]^n \to \hat{\pi}.$$

The theorems stated in this section should be taken only for what they are: preliminary and tentative endeavors to formulate a theory. Its shortcomings should be obvious to the reader and may stimulate him to try to remedy this. One would prefer to give the conditions in terms of the operators

$$M_p = \int_{-\infty}^{\infty} \left(\frac{\lambda}{n}\right)^p M(d\lambda)$$

rather than expressed in $M_{[p]}$. At present one knows how to do this only by means of complicated and practically unwieldy assumptions. We clearly need more experience in handling limit theorems on special groups before an adequate theory can be formulated.

Until then we can use results like those given above as a guide. The usual procedure will be first to determine the resolutions $E_g(\lambda)$ and then to compute the generalized resolutions $M(\lambda)$ and $N(\lambda)$. These are then used to form the relevant operators $M_1, M_{[1]}, N_1$ and so on. In this way we will get an indication of what limit theorems to expect.

5.5. Illustrations

5.5.1. It may be illuminating to start with an extremely simple case, the circle group $T^1 = \{\varphi \mid -\pi \leqslant \varphi < \pi\}$. Since the irreducible unitary representations $U(g)$ are then simply $e^{i\nu g}$, $\nu = 0, \pm 1, \pm 2, \ldots$ our $H(g)$ are then

$$H_0(g) \equiv 0$$
$$H_1(g) = g$$

$$H_2(g) = \begin{cases} 2g, & -\dfrac{\pi}{2} \leqslant g < \dfrac{\pi}{2} \\[2ex] 2g - 2\pi, & \dfrac{\pi}{2} \leqslant g < \pi \\[2ex] 2g + 2\pi, & -\pi \leqslant g < -\dfrac{\pi}{2} \end{cases}$$

$$H_3(g) = \begin{cases} 3g, & -\dfrac{\pi}{3} \leqslant g < \dfrac{\pi}{3} \\[2ex] 3g - 2\pi, & \dfrac{\pi}{3} \leqslant g < \dfrac{2\pi}{3} \\[2ex] 3g - 4\pi, & \dfrac{2\pi}{3} \leqslant g < \pi \\[2ex] 3g + 2\pi, & -\dfrac{2\pi}{3} \leqslant g < -\dfrac{\pi}{3} \\[2ex] 3g + 4\pi, & -\pi \leqslant g < -\dfrac{2\pi}{3} \end{cases}$$

     etc.

To exemplify Theorem 5.3.1. we may consider the rectangular distribution R_n on the intervals $(a/n, b/n)$. Then for any ν and for sufficiently large values of n

$$\int_G H_\nu(g) R_n(dg) = \frac{\nu}{n} \frac{a+b}{2} = \frac{1}{n} H_\nu\left(\frac{a+b}{2}\right)$$

and

$$\int_G H_\nu^2(g) R_n(dg) = \frac{\nu^2}{n^2} \frac{b^3 - a^3}{3(b-a)} = 0\left(\frac{1}{n^2}\right).$$

Then, of course, the theorem applies and $R_n^{n*} \to \delta_{(a+b)/2}$. Or take $b = -a$, so that $\bar{g} = e$, and large n. Then

$$\int_G H_\nu^2(g) R_{\sqrt{n}}(dg) = \frac{\nu^2 b^2}{3n}$$

and

$$\left\| \int_{-\pi}^{\pi} |\lambda|^3 M(d\lambda) \right\| = \int_G |H_\nu(g)|^3 R_{\sqrt{n}}(dg) = 0\left(\frac{1}{n^{3/2}}\right) = o\left(\frac{1}{n}\right)$$

so that we should expect $R_{\sqrt{n}}^{n*}$ to converge to the probability distribution π whose Fourier transform $\hat{\pi}$ is

$$\hat{\pi} = \exp -\frac{\nu^2 b^2}{6}.$$

It is easily seen that if x is a real valued normal variable with mean 0 and standard deviation σ, then $g \equiv x$ reduced to the interval $(-\pi, \pi)$ modulo 2π has the Fourier transform

$$\int_{-\pi}^{\pi} e^{i\nu g} P(dg) = \frac{1}{\sqrt{2\pi}\,\sigma} \int_{-\infty}^{\infty} e^{i\nu g - \frac{x^2}{2\sigma^2}} dx = e^{-\frac{\sigma^2 \nu^2}{2}}.$$

In the present case we have $\sigma = \dfrac{b}{\sqrt{3}}$. The limiting distribution is, of course, here simply the normal distribution wrapped around the circumference of the unit circle (see also 4.4.1).

Let us pause a moment and reflect on this example. When applying Theorem 5.3.1 we verify two conditions about moments of the first and second order. But in the present case the space $\mathcal{H}$, in which our unitary representations operate, is finite dimensional and this makes it possible to formulate the conditions in a simpler form. Write, in matrix language,

$$U(g) = \{u_{\nu\mu}(g), \nu, \mu = 1, 2, \ldots d\},$$

and assume that

$$\int_G u_{\nu\mu}(g) P_n(dg) = \delta_{\nu\mu} + \frac{a_{\nu\mu}}{n} + o\left(\frac{1}{n}\right), \ A = \{a_{\nu\mu}\}.$$

Then clearly

$$\lim_{n\to\infty} (\hat{P}_n)^n = \lim_{n\to\infty} \left[I - \frac{A}{n} + o\left(\frac{1}{n}\right)\right]^n = \exp A.$$

If $e^A = U(g_0)$ we get Theorem 5.3.1. and if $e^A = \hat{\pi}$ we get Theorem 5.3.2. In particular let us introduce the function

$$u(g) = \|U(g) - I\|$$

and demand that
$$\int_G u(g)\,P_n(dg) = o\!\left(\frac{1}{n}\right).$$

Then
$$\|\hat{P}_n - I\| \leqslant \int_G \|U(g) - I\|\,P_n(dg) = o\!\left(\frac{1}{n}\right)$$

so that $P_n^{n*} \to \delta_e$. We could also form a function built from all the representations simultaneously

$$v(g) = \sum_1^\infty k_\nu \|U_\nu(g) - I\|, \quad k_\nu > 0, \quad \sum k_\nu < \infty.$$

Then
$$\int v(g)\,P_n(dg) = o\!\left(\frac{1}{n}\right)$$

implies again that $P_n^{n*} \to \delta_e$.

5.5.2. A more interesting but also a bit more complicated stochastic group is the following which is neither compact nor commutative. Consider the group of linear transformations of the real line $x \to \alpha x + \beta$, where $\alpha > 0$ (see Notes 5.5.2).

Writing the group element $g = (\alpha, \beta)$ and defining the neighbourhoods in the obvious way this group, G, is locally compact. Its right invariant measure is given by

$$v(dg) = \frac{d\alpha\,d\beta}{\alpha}.$$

Its irreducible unitary transformations are known to consist of one-dimensional representations of the form α^{it}, t real, and certain infinite dimensional ones.

The one-dimensional representations will not be discussed here since they behave just as in the classical case. The infinite dimensional ones are given as follows. Let H^+ be the Hilbert space of functions $f(\lambda)$ defined on the positive real line $0 < \lambda < \infty$ and quadratically integrable with respect to Lebesgue measure on this half line; the inner product is defined as usual. Putting
$$U^+(g)f(\lambda) = e^{i\lambda\beta}f(\lambda\alpha)\sqrt{\alpha}$$
we have
$$\|U^+(g)f\|^2 = \int_0^\infty |f(\lambda\alpha)|^2 \alpha\,d\lambda = \int_0^\infty |f(\mu)|^2 d\mu = \|f\|^2,$$

and $U(g)$ is unitary. Further for $g_\nu = (\alpha_\nu, \beta_\nu), \nu = 1, 2$, we have

$$U(g_1)U(g_2)f(\lambda) = U(g_1)e^{i\lambda\beta_2}f(\lambda\alpha_2)\sqrt{\alpha_2}$$
$$= e^{i\lambda\beta_1 + i\lambda\alpha_1\beta_2}f(\lambda\alpha_1\alpha_2)\sqrt{\alpha_1\alpha_2} = U(g)f(\lambda),$$

with $g = g_1 g_2$ so that $U(g)$ is a representation; It is also clear that $U(g)$ is continuous. In the same way one introduces H^- and $U^-(g)$. It is known that the representations described form together the set of all irreducible and unitary representations.

Let us show that G has the P property. Consider the set

$$S = \left\{ g \left| \frac{1}{A} < \alpha < A, \; -B < \beta < B \right. \right\}, \quad A > 1,$$

and its indicator function $I(g) = I_s(g)$. Convolving $I(g)$ with $\check{I}(g)$ and dividing by $\nu(S)$ we get a normed positive definite function $\varphi(g)$

$$\varphi(g) = \frac{\displaystyle\int_G I(y) I(yg) \nu(dy)}{\nu(S)}$$

But the numerator is equal to

$$\nu\{y \,|\, y \in S, yg \in S\} = 2 \log A (1 + 0(1)) 2B(1 + 0(1)) = \nu(S)(1 + o(1))$$

if $A = o(B)$ as A and B tend to infinity. This means that the positive definite function $\varphi(g)$ approximates the constant function 1 in the required way and the assertion is proved.

This group has unique nth roots. Actually, for any real t we have $g^t = (\alpha_t, \beta_t)$ with

$$\left. \begin{aligned} \alpha_t &= \alpha^t \\ \beta_t &= \beta \frac{1 - \alpha^t}{1 - \alpha} \end{aligned} \right\}$$

for $\alpha \neq 1$ and

$$\left. \begin{aligned} \alpha_t &= 1 \\ \beta_t &= t\beta \end{aligned} \right\}$$

for $\alpha = 1$. Studying the operator $U(g^t) = \exp itH(g)$ for small values of the parameter t we find

$$iH(g)f(\lambda) = \left[i\lambda\beta \frac{\log \alpha}{\alpha - 1} + \frac{\log \alpha}{2} \right] f(\lambda) + \lambda \log \alpha f'(\lambda)$$

for $f \in \mathcal{D}$, say the set of all functions $f(\lambda)$ vanishing outside of finite intervals and with a continuous derivative.

Consider a distribution P over G such that the following two integrals converge

$$\left.\begin{array}{l} \displaystyle\int_G \log \alpha \, P(dg) = a \\[2ex] \displaystyle\int_G \beta\frac{\log \alpha}{\alpha - 1} P(dg) = b \end{array}\right\}$$

and in the following we assume for simplicity that α and β are independent, although the latter condition is not very essential for the problem. Then

$$i\int_G H(g)\, P(dg) = ibM + \frac{a}{2}\,I + aMD,$$

where M is the operator consisting of multiplication by λ and D is the differentiation operator. Then

$$i\int_G H(g) P(dg) = iH(\bar{g})$$

if we choose $\bar{g}$, the average, as $\bar{g} = (\xi, \eta)$

$$\left.\begin{array}{l} \xi = e^a \\[2ex] \eta = b\cdot\dfrac{e^a - 1}{a} \end{array}\right\}$$

for $a \neq 0$ and as

$$\left.\begin{array}{l} \xi = 1 \\[1ex] \eta = b \end{array}\right\}$$

for $a = 0$.

We would therefore expect the stochastic group elements $\gamma_n = g_1^{1/n}\, g_2^{1/n}\ldots g_n^{1/n}$ to converge in probability to the average $\bar{g}$. Because of the fairly simple structure of the group we can show this directly. We have, putting $\gamma_n = (\alpha^{(n)}, \beta^{(n)})$

$$\left\{\begin{array}{l} \alpha^{(n)} = (\alpha_1 \alpha_2 \ldots \alpha_n)^{1/n} \\[2ex] \beta^{(n)} = \beta_1 \dfrac{1 - \alpha_1^{1/n}}{1 - \alpha_1} + \beta_2 \dfrac{1 - \alpha_2^{1/n}}{1 - \alpha_2}\, \alpha_1^{1/n} + \ldots + \beta_n \dfrac{1 - \alpha_n^{1/n}}{1 - \alpha_n}\, \alpha_1^{1/n}\, \alpha_2^{1/n}\ldots\alpha_{n-1}^{1/n}. \end{array}\right.$$

Taking logarithms, the law of large numbers tells us immediately that $\alpha^{(n)} \to \xi = e^a = e^{E \log \alpha}$ in probability. Writing

$$\left\{\begin{array}{l} b_1^{(n)} = \displaystyle\sum_1^n \beta_\nu \dfrac{1 - \alpha_\nu^{1/n}}{1 - \alpha_\nu} \\[3ex] b_2^{(n)} = \dfrac{1}{n}\displaystyle\sum_1^n \beta_\nu \dfrac{\log \alpha_\nu}{\alpha_\nu - 1} \end{array}\right.$$

we see that $b_2^{(n)} \to b$ in probability. But

$$b_1^{(n)} - b_2^{(n)} = \sum \beta_\nu \frac{e^{1/n \log \alpha_\nu} - 1 - \frac{1}{n} \log \alpha_\nu}{\alpha_\nu - 1}$$

so that it follows from the independence of α_ν and β_ν and by an elementary but cumbersome dominated convergence argument

$$E\left|b_1^{(n)} - b_2^{(n)}\right| \leq nE|\beta| E\left|\frac{e^{1/n \log \alpha} - 1 - \frac{1}{n} \log \alpha}{\alpha - 1}\right| \to 0$$

and $b_1^{(n)} \to b$ in probability. To complete the argument we split up the sum defining $\beta^{(n)}$ into many long blocks, such that the factors of form $(\alpha_1 \alpha_2 \ldots \alpha_\nu)^{1/\nu}$ are nearly constant, $\sim e^{\frac{\nu}{n} a}$, and apply the above reasoning. This proves the convergence.

Now let us go on to examine the distribution of $\delta_n = g_1^{1/\sqrt{n}} g_2^{1/\sqrt{n}} \ldots g_n^{1/\sqrt{n}}$ when $a = 0$, $b = 0$, i.e. the average $\bar{g} = e$. We must now also require that the integrals

$$\begin{cases} E (\log \alpha)^2 = 1 \\ E \beta^2 \left(\frac{\log \alpha}{\alpha - 1}\right)^2 = c \end{cases}$$

exist; the norming of the first integral is arbitrary and not essential. We then have

$$-H^2(g) = \left\{ i\lambda\beta \frac{\log \alpha}{\alpha - 1} + \frac{\log \alpha}{2} + \lambda \log \alpha D \right\} \left\{ i\lambda\beta \frac{\log \alpha}{\alpha - 1} \right.$$

$$\left. + \frac{\log \alpha}{2} + \lambda \log \alpha D \right\} f(\lambda) = A(\alpha, \beta) f(\lambda) + B(\alpha, \beta) f'(\lambda) + C(\alpha, \beta) f''(\lambda)$$

with

$$\begin{cases} A(\alpha, \beta) = (\log \alpha)^2 \left[-\frac{\lambda^2 \beta^2}{(\alpha - 1)^2} + 2i \frac{\lambda\beta}{\alpha - 1} + \frac{1}{4} \right] \\ B(\alpha, \beta) = (\log \alpha)^2 \left[\frac{2i\lambda^2\beta}{\alpha - 1} + 2\lambda \right] \\ C(\alpha, \beta) = \lambda^2 (\log \alpha)^2. \end{cases}$$

Hence

$$-M_2 = -\int_G H^2(g) P(dg) = A + B \cdot D + C \cdot D^2,$$

where we have introduced the polynomials

$$\begin{cases} A = a_2\,\lambda^2 + a_1\,\lambda + a_0 \\ B = b_2\,\lambda^2 + b_1\,\lambda \\ C = c_2\,\lambda^2 \end{cases}$$

with the coefficients

$$\begin{cases} a_0 = \tfrac{1}{4}\,E\,(\log\alpha)^2 \\[2mm] a_1 = 2\,i\,E\,(\log\alpha)^2\,\dfrac{\beta}{\alpha-1} \\[2mm] a_2 = -E\,(\log\alpha)^2\left(\dfrac{\beta}{\alpha-1}\right)^2 \\[2mm] b_1 = 2\,E\,(\log\alpha)^2 \\[2mm] b_2 = 2\,i\,E\,(\log\alpha)^2\,\dfrac{\beta}{\alpha-1} \\[2mm] c_2 = E\,(\log\alpha)^2. \end{cases}$$

With our choice of constants

$$\begin{cases} A = -c\,\lambda^2 + \tfrac{1}{4} \\ B = 2\,\lambda \\ C = \lambda^2. \end{cases}$$

For a sufficiently well-behaved $f(\lambda)$ we put

$$f(\lambda, t) = \exp\left(-\frac{t}{2}\,M_2\right) f(\lambda)$$

satisfying the parabolic equation

$$\frac{\partial f}{\partial t} = \frac{1}{8} f - \frac{c}{2}\lambda^2 f + \frac{1}{2}(\lambda^2 f')'.$$

Then the operator $f(\lambda) \to f(\lambda, 1)$ should be expressed in terms of the unitary representations to give the required limit theorem. We shall return to this in Chapter 7.

Before we leave this group let us consider the following limit problem that differs from the ones above. We consider now the element $\delta_n = (\alpha^{(n)}, \beta^{(n)}) = (g_1 g_2 \ldots g_n)^{1/n}$. We have

$$\begin{cases} \alpha^{(n)} = (\alpha_1 \alpha_2 \ldots \alpha_n)^{1/n} \\ \beta^{(n)} = [\beta_1 + \beta_2 \alpha_1 + \beta_3 \alpha_1 \alpha_2 + \ldots + \beta_n \alpha_1 \alpha_2 \ldots \alpha_{n-1}] \dfrac{1 - (\alpha_1 \alpha_2 \ldots \alpha_n)^{1/n}}{1 - \alpha_1 \alpha_2 \ldots \alpha_n}. \end{cases}$$

For the first coordinate $\alpha^{(n)}$ nothing new happens, it converges in probability to the geometric mean $\xi = \exp E \log \alpha$ which we assume exists. To study the second coordinate take first the case $\xi < 1$. Then $\beta^{(n)}$ converges distribution-wise (and in probability) and the limit distribution is that of the stochastic variable

$$\beta^* = [\beta_1 + \beta_2 \alpha_1 + \beta_3 \alpha_1 \alpha_2 + \ldots](1 - \xi),$$

where the infinite series converges almost certainly if $E\beta_\nu$ exists since, with probability one $(\alpha_1 \alpha_2 \ldots \alpha_n)^{1/n} \to \xi$, so that the factors $\alpha_1 \alpha_2 \ldots \alpha_n$ behave asymptotically as ξ^n, $\xi < 1$. On the other hand if $\xi > 1$ then $\beta^{(n)}$ still converges distributionwise (but not in probability), but now to the distribution of the stochastic variable

$$\beta^{**} = \left[\frac{\beta_1}{\alpha_1} + \frac{\beta_2}{\alpha_1 \alpha_2} + \frac{\beta_3}{\alpha_1 \alpha_2 \alpha_3} + \ldots \right](\xi - 1).$$

Hence in both these cases we have a non-degenerate limit distribution for δ_n when we use the norming $(g_1 g_2 \ldots g_n)^{1/n}$. Compare this with what happens for the norming $\gamma_n = g_1^{1/n} g_2^{1/n} \ldots g_n^{1/n}$ and with the classical law of large numbers, when the limits are degenerate.

5.5.3. Now let us discuss the spectrum of the operator T associated with countable stochastic groups. Let G be a countable group and introduce the probability p_{gh} of going from g to h via right multiplication, so that $p_{gh} = P(g^{-1}h)$. These transition probabilities form a matrix

$$M = \{p_{gh}; \ g \text{ and } h \in G\},$$

and it will be assumed that M is symmetric. M can be considered as an operator in the l_2-space of quadratically summable sequences $f_g, g \in G$. It is clear that M is identical with the operator T (see section 3.1).

$$Tf_g = \int_G f_{gh} P(dh) = \sum_{h \in G} f_{gh} P(h) = \sum_{k \in G} P(g^{-1}k) f_k = \sum_{k \in G} p_{gk} f_k = (Mf)_g.$$

The spectrum of M is real (symmetric measure) and contained in the interval $(-1, 1)$. The spectral radius $r = \max |\lambda|$, $\lambda \in$ spectrum, is of particular interest, and especially one wants to know whether $r < 1$ or $r = 1$. See also the related discussion in section 3.1. In general it is difficult to determine r. It is of interest that the determination is possible in the following case.

Let G be a *free group* generated by h free generators. If P has a uniform symmetric distribution on these generators then (see Notes 5.5.3.)

$$r = \sqrt{\frac{2h-1}{h^2}}.$$

This is less than 1 if $h \geqslant 2$. On such groups it is not possible to approximate the constant 1 in the way described earlier in this chapter by functions of the form $\varphi * \tilde{\varphi}$.

But finite groups or commutative countable groups have $r = 1$.

CHAPTER 6

STOCHASTIC LINEAR SPACES

6.1. Probabilities on a Banach space

In the last chapter we tried to formulate limit theorems similar to the law of large numbers and the central limit theorem. This was possible to some degree but it is clear that we can expect much more informative answers if we impose more detailed structure on the group. In particular, if we demand that the group and the operation corresponding to taking powers behave in a simple way, then progress will become much easier.

The concept we have in mind is that of a *linear space*, in particular a *Banach space*. It is of course commutative and the operation just mentioned, which is now just scalar multiplication, is distributive. At first glance one may think that this is of so much help to us that we should get the desired results without any more ado by applying some straightforward Fourier analysis. This is not so and the reason for that is that we now have to deal with spaces that are not locally compact. We no longer have any invariant measure, and the infinite-dimensional nature of the space has some not very agreeable consequences that we will meet below.

Consider a Banach space X (we will usually take it as real) with elements x and a norm $\|x\|$. The dual space consisting of all the continuous linear functionals $x^*(x)$ is denoted by X^*. We shall assume throughout that both X and X^* are separable. The reader should observe that some of the statements made in this chapter hold (or are known to hold) only under a separability condition.

Among the particular Banach spaces we will of course pay special attention to the separable *Hilbert* space.

We shall consider probability measures P on X defined on the σ-algebra generated by the open subsets of X. Then any continuous linear functional $x^*(x)$ is measurable. For the actual construction of the measure it may be more convenient to start from the finite dimensional distribution of $x_1^*(x)$, $x_2^*(x),...,x_n^*(x)$, with $x_\nu^* \in X^*$, i.e. we start with the set algebra generated by cylinder sets of the form

$$\{x \,|\, (x_1^*(x), x_2^*(x), \ldots, x_n^*(x)) \in B\},$$

where B is an arbitrary Borel set in R^n. We then apply Kolmogorov's extension theorem to extend the probability measure to the corresponding σ-algebra $\mathcal{L}$, which contains all the open sets (see Notes 6.1.$_1$). This alternative approach would start from the assumption that $x^*(x)$ is P-measurable for any $x^* \in X^*$; such a stochastic element is called L-stochastic.

Here it is appropriate to mention the concept of a *family of unicity*. Let $\mathcal{U} = \{U\}$ be a family of sets (say Borel sets) such that knowledge of the values $P(U)$ of a probability measure P on all the U's determines P completely. Then $\mathcal{U}$ is called a family of unicity. In the present case all the half-spaces, $\{x \,|\, x^*(x) < c\}$, form a family of unicity. This has been known for a long time for finite dimensional Euclidean spaces (see Notes 6.1.$_2$).

On a Banach space the *mean value* m of a probability distribution P is defined via a Pettis integral. We shall say that P has the mean value m if $x^*(x)$ is integrable for any x^* and if there is an element m satisfying

$$x^*(m) = \int_X x^*(x)\, P(dx), \text{ all } x^* \in X^*.$$

If m exists it is unique and we will denote it by $m = Ex$. Further it is clear that

1) If Ex_1 and Ex_2 exist, then $E(x_1 + x_2)$ exists and is the sum of Ex_1 and Ex_2.

2) If x is a constant element x_0 with probability one then Ex exists and is equal to x_0.

3) If Ex exists, then Ecx exists for any constant scalar c and is equal to $c \cdot Ex$.

4) If Ex exists and L is a bounded linear operator from X to another Banach space Y, then $E[L(x)]$ exists and is equal to $L[E(x)]$.

5) If $E\|x\| < \infty$ then Ex exists and $\|Ex\| \leqslant E\|x\|$.

These properties are proved as for Pettis integrals (see Notes 6.1.$_3$).

If X is a Hilbert space we can also introduce a linear operator Q playing a role similar to that of a second order moment matrix for finite dimensional Euclidean spaces. If $E\|x\|^2 < \infty$ we define Q through the equation

$$(Qz, z) = E(x, z)^2$$

that is $\qquad\qquad (Qz, y) = E(x, z)(x, y),$

and similarly a *covariance operator* C through

$$(Cz, z) = E(x - Ex, z)^2.$$

Q is Hermitian, non-negative, bounded and, if e_ν are unit vectors along the coordinate axes, since $(Qy, z) = E(x, y)(x, z)$, we have

$$\sum_{\nu, \mu} [(Qe_\nu, e_\mu)]^2 = \sum_{\nu, \mu} [E(x, e_\nu)(x, e_\mu)]^2 \leqslant (\sum_\nu E[x, e_\nu)]^2)^2 = (E\|x\|^2)^2 < \infty$$

so that Q is a Hilbert-Schmidt operator and hence completely continuous. Further

$$\sum_\nu (Qe_\nu, e_\nu) = E\|x\|^2 < \infty$$

so that Q has finite trace. The same holds for C. Such operators are called S-operators. To a sum $x_1 + x_2$ of two stochastically independent elements in X corresponds a covariance operator which is just the sum of the covariance operators for x_1 and x_2.

If X is represented as the sequence space l_2 then Q and S are the usual infinite dimensional moment matrices.

When considering a sequence of stochastic elements $x_1, x_2, \ldots,$ in X we have several useful notions of convergence. We may have almost certain weak convergence, $P\{x^*(x_\nu) \to x^*(x), \text{ all } x^* \in X^*\} = 1$, and almost certain strong convergence, $P\{\|x_n - x\| \to 0\} = 1$. We may have convergence in probability, $P\{\|x_n - x\| > \varepsilon\} \to 0$ for every positive ε. Finally we can have convergence in the mean with index α, $E\|x_n - x\|^\alpha \to 0$.

As far as convergence of probability distributions $P_n \to P$ is concerned the most important notion is perhaps, as before, weak convergence meaning that

$$\int_X f(x) P_n(dx) \to \int_X f(x) P(dx)$$

for every continuous and bounded function $f(x)$. This is equivalent to demanding that the distribution functions

$$P_n\{x \mid g(x) \leqslant c\} \to P\{x \mid g(x) \leqslant c\}$$

for every continuous $g(x)$ for every c where the right member is continuous. It can be shown that this happens if $P_n(U)$ converges on a family $\{U\}$ of unicity and the sequence $\{P_n\}$ is compact.

It may be mentioned that $\mathcal{P}(X)$ can be made into a complete, separable metric space, where the following metric corresponds to weak convergence. Take an arbitrary closed set C and form the open set

$$C^\varepsilon = \{x \mid \inf_{y \in C} \|x - y\| < \varepsilon\}.$$

Define ε_1 as the greatest lower bound of ε's satisfying

$$P_2(C) < P_1(C^\varepsilon) + \varepsilon.$$

Similarly ε_2 is the greatest lower bound of ε's satisfying

$$P_1(C) < P_2(C^\varepsilon) + \varepsilon.$$

The metric is then defined through

$$\varrho(P_1, P_2) = \max(\varepsilon_1, \varepsilon_2).$$

See Notes $6.1._4$.

6.2. Fourier analysis in a stochastic Banach space

Let X be a real Banach space and P a probability distribution defined on the Borel sets of X. It is then natural to define the Fourier transform or *characteristic functional* of P as follows.

Definition 6.2. The characteristic function (or functional) $\hat{P}(x^*)$ is defined as the function on X^*

$$\hat{P}(x^*) = E \exp i x^*(x) = \int_X \exp i x^*(x) P(dx).$$

Some of the basic properties of characteristic functionals are easily established.

Theorem 6.2.1. On $\mathcal{P}(X)$ the following holds:

1) $\hat{P}(x^*)$ is uniformly strongly continuous in x^* as well as weakly continuous.

2) $\hat{P}(0) = 1$

3) $\widehat{P_1 * P_2}(x^*) = \hat{P}_1(x^*) \cdot \hat{P}_2(x^*)$

4) $\hat{P}(x^*)$ is a positive definite function on X^*.

5) If $A_{p+1} = \int_X \|x\|^{p+1} P(dx) < \infty$, then $\hat{P}(x^*)$

$$= \sum_{\nu=0}^{p} \frac{i^p}{p!} E(x^*(x))^p + \frac{\theta}{(p+1)!} A_{p+1}, \quad |\theta| \leqslant 1.$$

6) $\hat{P}(x^*)$ determines P uniquely

7) If a sequence P_n converges weakly to a probability distribution P, then $\hat{P}_n(x^*) \to \hat{P}(x^*)$ for all $x^* \in X^*$.

Proof: The statements 2), 3), 4) and 5) are proved exactly as on the real line and 7) follows directly from the definition of weak convergence.

To prove 1) we have

$$\left|\hat{P}(x_1^*) - \hat{P}(x_2^*)\right| \leqslant E\left|e^{ix_1^*(x)} - e^{ix_2^*(x)}\right| = E\left|1 - e^{iz^*(x)}\right|$$

with $z^* = x_1^* - x_2^*$. Choose a set S with $P(S) > 1 - \delta$ and of bounded elements, $\sup_{x \in S} \|x\| = M < \infty$.

$$\left|\hat{P}(x_1^*) - \hat{P}(x_2^*)\right| \leqslant \int_S \left|1 - e^{iz^*(x)}\right| P(dx) + 2\delta \leqslant 3\delta$$

if we take z^* so near 0 that

$$\left|1 - e^{iz^*(x)}\right| \leqslant \delta$$

for all $x \in S$. This proves the first part of 1. The second part is verified in the same way since weak convergence $x_2^* \to x_1^*$ in X^* implies $x_2^*(x) - x_1^*(x) \to 0$ for every $x \in X$ and Lebesgue's theorem on bounded convergence applies.

Statement 6) holds since the probability mass in any set of the form $U = \{x \mid x^*(x) \leqslant c\}$ can be computed from $\hat{P}(x^*)$ just as on the real line. But the U's form a family of unicity and hence the whole of P is uniquely determined.

Before going ahead let us note the simple fact that we need not look for any idempotent measures in a linear space since there are none (except the trivial one δ_0). Indeed if $P \in \mathcal{P}(X)$ is non-trivial there must be some $x^* \in X^*$ such that $P\{x^*(x) \neq 0\} > 0$. Consider the Fourier transform

$$\hat{P}(tx^*) = E \exp tx^*(x)$$

for real values of t. Since P is idempotent $(\hat{P})^2 = \hat{P}$ so that $\hat{P}(tx^*)$ can take only the values 1 or 0. But it is a continuous function of t which is equal to one for $t = 0$. Hence the characteristic function of the stochastic variable $x^*(x)$ is identically one so that $x^*(x) \equiv 0$ contrary to the assumption.

Now we meet an essential difficulty. We would like to assert, of course, that $\hat{P}_n(x^*) \to \hat{P}(x^*)$ implies $P_n \to P$ weakly. That this is not so in general is seen from the following example. Let $x_1, x_2, x_3, \ldots$ be a sequence of elements in X converging weakly but not strongly to an element x_0. Let P_n assign the measure 1 to the element x_n. Then $x^*(x_n) \to x^*(x_0)$ and $\exp ix^*(x_n) = \hat{P}_n(x^*) \to \exp ix^*(x_0) = P_0(x^*)$ for any $x^* \in X^*$. On the other hand, let $f(x)$ be a continuous and bounded function $= 1$ in $x = x_0$ and vanishing outside a sphere $S_\varepsilon = \{x \mid \|x - x_0\| < \varepsilon\}$ so that for a small ε and some arbitrarily large values of n

$$\int_X f(x)\, P_n(dx) = 0 \longmapsto 1 = \int_X f(x)\, P_0(dx).$$

If X is a Hilbert space l_2 we may choose the element x_n as a unit vector on the nth coordinate axis; this brings forwards very clearly that the difficulty is caused by the infinite dimensional behavior of the space. To remedy this we observe the following (see Notes 6.2.$_1$).

Any $P \in \mathcal{P}(X)$ is a *compact measure*: to any $\varepsilon > 0$ there is a compact set $C \subset X$ such that $P(C) \geqslant 1 - \varepsilon$. Indeed, let us choose a sequence $x_1, x_2, x_3, \ldots$ dense in X and form the spheres $S_n^k = \{x \mid \|x - x_n\| \leqslant 1/k\}$ with the points x_n as centers and of radii $1/k$. Choose m_1 so large that

$$P\left\{ \bigcup_{n=1}^{m_1} S_n^1 \right\} > 1 - \varepsilon.$$

Then choose m_2 so large that

$$P\left\{ \left(\bigcup_{n=1}^{m_1} S_n^1 \right) \cap \left(\bigcup_{n=1}^{m_2} S_n^2 \right) \right\} > 1 - \varepsilon,$$

and so on. The set
$$C = \bigcap_{i=1}^{\infty} \left(\bigcup_{n=1}^{m_i} S_n^i \right)$$

is compact and has measure $P(C) \geqslant 1 - \varepsilon$.

Now we introduce the

Definition 6.2.2. The probability measures $P_n \in \mathcal{P}(S)$ are called uniformly compact if for any positive ε there is a compact set $C \subset X$ such that $P_n(C) \geqslant 1 - \varepsilon$ for all n.

We then have

Theorem 6.2.2. In order that $P_n \to P$ weakly it is necessary and sufficient that the P_n are uniformly compact and that $\hat{P}_n(x^*) \to \hat{P}(x^*)$ for all $x^* \in X^*$.

The proof is left to the reader. It goes without saying that the uniform compactness assumption in the above theorem may be difficult to verify, so that the theorem is of limited usefulness. For a Hilbert space $X = l_2$ the following is a sufficient condition (see Notes 6.2.$_2$). Write $x = (x_1, x_2, \ldots)$ and introduce the quantity

$$R_N = \sup_n E\left[\sum_N^{\infty} x_\nu^2 \right].$$

Note that the supremum is taken over the P_n's.

Theorem 6.2.3. If $P_n \in \mathcal{P}(l_2)$ satisfy

1) $$\lim_N R_N = 0,$$

2) $$\hat{P}_n(x^*) \to \hat{P}(x^*), \text{ all } x^* \in X^*,$$

then $P_n \to P$ weakly.

Proof: Consider the compact set

$$C = \bigcap_{k=1}^{\infty} E_k,$$

where

$$E_k = \left\{ x \;\middle|\; \sum_{N_k}^{\infty} x_\nu^2 \leq \frac{1}{l_k} \right\}.$$

Here we have choosen an increasing sequence of integers $N_k \uparrow + \infty$ with $N_1 = 1$ and positive $l_k \to \infty$ such that

$$\sum_{1}^{\infty} l_k R_{N_k} < \varepsilon.$$

But then

$$1 - P_n(C) = P_n(C^*) \leq \sum_{k=1}^{\infty} P_n(E_k^*) \leq \sum_{k=1}^{\infty} l_k R_{N_k} < \varepsilon$$

and Theorem 6.2.2. applies.

The next difficulty arises when we try to invert statement 4) of Theorem 6.2.1. The inverse of the statement is not true without further qualifications. Indeed consider the function

$$\varphi(x^*) = \exp - \tfrac{1}{2} \|x^*\|^2, \; x^* \in l_2.$$

This function is continuous in the strong topology and positive definite (just as normal characteristic functions in finite dimensional spaces are positive definite). If φ corresponds to a measure P on l_2 then the distributions of the coordinates $x_\nu = (x, e_\nu)$ must obviously be independent stochastic variables $N(0,1)$. The stochastic variables

$$R_n^2 = x_1^2 + x_2^2 + \ldots + x_n^2$$

obviously converge to $+\infty$ in probability as $n \to \infty$. But then $P\{\|x\| \leq R\} = 0$ for any R, which is impossible. One might guess that the trouble is the sort of continuity used. Replacing strong convergence by weak convergence does not lead to success as can be shown by a modification of the above example.

The correct mode of convergence is the one corresponding to the neighborhoods

$$(Sx^*, x^*) < 1$$

for arbitrary S-operators. Let us call this the S-topology (see Notes $6.2._3$).

Theorem 6.2.4. Let $\varphi(x^*)$ be a complex-valued function of x^* on the Hilbert space X^* with $\varphi(0)=1$. In order that $\varphi(x^*)$ be the Fourier transform $\hat{P}(x^*)$ of some $P \in \mathcal{P}(X)$ it is necessary and sufficient that φ is positive definite and continuous in the S-topology at $x^*=0$.

Proof of necessity: For $P \in \mathcal{P}(X)$ we choose a compact set $C \subset X$ such that $P(C) > 1 - \varepsilon$. Then

$$|1 - \hat{P}(x^*)| \leqslant 2\varepsilon + \int_C [1 - \cos(x^*, x)] P(dx)$$

$$+ \int_C |\sin(x^*, x)| P(dx) \leqslant 2\varepsilon + \frac{1}{2} \int_C (x^*, x)^2 P(dx) + \sqrt{\int_C (x^*, x)^2 P(dx)}.$$

Introduce S-operators through

$$(S_\varepsilon x^*, x^*) = \frac{1}{\varepsilon} \int_C (x^*, x)^2 P(dx),$$

then the assertion follows.

Proof of sufficiency: Consider the space Y of *all* sequences $y = (y_1, y_2, \ldots)$. Knowledge of $\varphi(x^*)$ determines a probability distribution P_n of any finite dimensional vector $(y_1, y_2, \ldots, y_n)$ and as usual we can extend this probability measure to the generated σ-algebra in Y. Now for any positive ε we can find an S operator (written in diagonal form in l_2)

$$(Sx^*, x^*) = \sum_1^\infty a_\nu (x_\nu^*)^2, \ \sum_1^\infty a_\nu < \infty, \ a_\nu \geqslant 0,$$

such that $|1 - \varphi(x^*)| \leqslant \varepsilon$ for x^* in $(Sx^*, x^*) < 1$. Then

$$P\left(\sum_1^n y_\nu^2 \geqslant 1\right) \leqslant \frac{\sqrt{e}}{\sqrt{e}-1} \int_{R^n} \left(1 - \exp\left[-\frac{1}{2}\sum_1^n y_\nu^2\right]\right) P_n(dy)$$

since
$$\frac{\sqrt{e}}{\sqrt{e}-1}(1 - e^{-\frac{1}{2}u}) \geqslant 1$$

for $|u| > 1$. Hence

$$P\left(\sum_1^n y_\nu^2 \geqslant 1\right) \leqslant \frac{\sqrt{e}}{\sqrt{e}-1}\frac{1}{(2\pi)^{n/2}}\int_{R^n}\{1-\varphi(y^*)\}\cdot\exp\left[-\frac{1}{2}\sum_1^n y_\nu^{*2}\right]dy^*$$

(use the properties of the normal frequency function in R^n and its characteristic function), and

$$P\left(\sum_1^n y_\nu^2 \geqslant 1\right) \leqslant \frac{\sqrt{e}}{\sqrt{e}-1}\varepsilon + \frac{\sqrt{e}}{\sqrt{e}-1}\frac{1}{(2\pi)^{n/2}}\cdot$$

$$\int_{\sum_1^n a_\nu(y_\nu^*)^2>1} 2\exp\left[-\frac{1}{2}\sum_1^n y_\nu^{*2}\right]dy^* = \frac{\sqrt{e}}{\sqrt{e}-1}\left(\varepsilon + 2\sum_1^n a_\nu\right),$$

since

$$\frac{1}{(2\pi)^{n/2}}\int_{\sum_1^n a_\nu(y_\nu^*)^2>1}\exp\left[-\frac{1}{2}\sum_1^n (y_\nu^*)^2\right]dy^* \leqslant \frac{1}{(2\pi)^{n/2}}\int_{R^n}\left[\sum_1^n a_\nu(y_\nu^*)^2\right]$$

$$\exp\left[-\frac{1}{2}\sum_1^n (y_\nu^*)^2\right]dy^* = \sum_1^n a_\nu.$$

Similarly,
$$P\left(\sum_1^n y_\nu^2 \geqslant C\right) \leqslant \frac{\sqrt{e}}{\sqrt{e}-1}\left(\varepsilon + 2\frac{\sum_1^n a_\nu}{C}\right).$$

But this implies that the probability measure P introduced in Y has all its mass in l_2, $P\left\{y \,\middle|\, \sum_1^\infty y_\nu^2 < \infty\right\} = 1$. The σ-algebra mentioned above can be

seen to contain all the open subsets of l_2, compare with the reasoning in Notes 6.1. Finally the Fourier transform of P in l_2 coincides with $\varphi(x^*)$ for all $x^* \in X^*$. This completes the proof.

See Notes 6.2.$_3$ for references.

In other Banach spaces the situation is, of course, more complicated. It will be instructive to return for a moment to the real line and we take as our point of departure the following elementary

Proposition Let $\varphi(t)$, $-\infty < t < \infty$, be a measurable and positive definite function. If

$$\varphi(0) = \lim_{\varepsilon\downarrow 0}\frac{1}{2\varepsilon}\int_{-\varepsilon}^\varepsilon \varphi(t)\,dt,$$

then $\varphi(t)$ is continuous.

Proof: It is well know that $\varphi(t)$ coincides almost everywhere with a continuous positive definite function $\varphi_c(t)$, $\varphi(t) = \varphi_c(t) + \varphi_0(t)$ with $\varphi_0(t) = 0$ a.e. Introduce for arbitrary t_ν, c_ν the functions

$$\left.\begin{aligned}
\Phi(t) &= \sum_{\nu,\mu} c_\nu \bar{c}_\mu \varphi(t + t_\nu - t_\mu) \\
\Phi_c(t) &= \sum_{\nu,\mu} c_\nu \bar{c}_\mu \varphi_c(t + t_\nu - t_\mu)
\end{aligned}\right\}.$$

Then $\Phi(t)$ is positive definite since for any s_k, a_k we have

$$\sum_{k,l} a_k \bar{a}_l \Phi(s_k - s_l) = \sum_{k,l,\nu,\mu} a_k c_\nu \overline{a_l c_\mu} \varphi(s_k + t_\nu - s_l - t_\mu)$$

which is non-negative by definition. This implies that $|\Phi(t)| \leqslant \Phi(0)$ so that

$$\Phi(0) \geqslant \lim_{\varepsilon \downarrow 0} \frac{1}{2\varepsilon} \int_{-\varepsilon}^{\varepsilon} \Phi(t)\,dt = \lim_{\varepsilon \downarrow 0} \frac{1}{2\varepsilon} \int_{-\varepsilon}^{\varepsilon} \Phi_0(t)\,dt + \Phi_c(0),$$

that is

$$\Phi(0) - \Phi_c(0) = \sum_{\nu,\mu} c_\nu \bar{c}_\mu \varphi_0(t_\nu - t_\mu) \geqslant 0$$

so that $\varphi_0(t)$ is positive definite. But we have assumed that

$$\varphi_0(0) = \varphi(0) - \varphi_c(0) = \lim_{\varepsilon \downarrow 0} \frac{1}{2\varepsilon} \int_{-\varepsilon}^{\varepsilon} \varphi(t)\,dt - \varphi_c(0)$$

$$= \lim_{\varepsilon \downarrow 0} \frac{1}{2\varepsilon} \int_{-\varepsilon}^{\varepsilon} \varphi_c(t)\,dt - \varphi_c(0) = 0.$$

Hence $\varphi_0(0) = 0$ which implies $\varphi_0(t) \equiv 0$.—It is clear that the proposition is valid for other singular integral kernels.

Let us now return to the Banach spaces and consider a simple example. Let $X = l_1$ consisting of summable sequences $x = (x_1, x_2, \ldots)$, $\|x\| = \sum_{1}^{\infty} |x_\nu| < \infty$, with the dual space $X^* = m$ consisting of bounded sequences $t = (t_1, t_2, \ldots)$, $\|t\| = \sup |t_\nu|$. Let the stochastic variables x_ν be independent with

$$\left\{\begin{aligned}
& P(x_\nu = 0) = \frac{1}{2} \\
& P\left(x_\nu = \frac{1}{\nu}\right) = P\left(x_\nu = -\frac{1}{\nu}\right) = \frac{1}{4}.
\end{aligned}\right.$$

This does not define a legitimate distribution in l_1 since

$$\sum_{1}^{\infty} E|x_\nu| = \sum_{1}^{\infty} \frac{1}{2\nu} = +\infty$$

so that $\sum_1^\infty |x_\nu|$ does not converge almost certainly (note that the x_ν are uniformly bounded). On the other hand, the "Fourier transform"

$$\varphi(t) = \varphi(t_1, t_2, \ldots) = E \, \exp \, i \sum t_\nu x_\nu = \prod_{\nu=1}^\infty \left[\frac{1}{2} + \frac{1}{2} \cos \frac{t_\nu}{\nu} \right]$$

converges for every $t \in m$. If $\|t\| \to 0$ then $\varphi(t) \to 1$. This shows that for the positive definite function $\varphi(t)$ strong continuity does not ensure that it is a Fourier transform in the strict sense that we have used.

We are now ready to formulate the continuity condition. Let Λ be a directed set of probability measures λ in X^* converging to δ_0 in a manner that will be made precise below. We shall say that $\varphi(t)$ is λ-continuous if

$$\lim_\Lambda \int_{X^*} \varphi(t) \lambda(dt) = \varphi(0).$$

Theorem. Let $\varphi(t)$ be a positive definite function on $X^* = m$ with $\varphi(0) = 1$. It is the Fourier transform of a probability measure on $X = l_1$ if and only if it is λ-continuous for λ-measures of the form

$$\lambda(dt) = \prod_1^n \frac{1}{\pi \varepsilon_\nu \left[1 + \dfrac{t_\nu^2}{\varepsilon_\nu^2} \right]} dt_1 \ldots dt_n \, \delta_0(dt_{n+1}) \ldots$$

$$\max_{1 \leqslant \nu \leqslant n} |\varepsilon_\nu| \to 0.$$

Proof: Supposing λ-continuity it follows that $\varphi(t_1, t_2, \ldots t_n, 0, 0, \ldots)$ is continuous in R^n (see the proof of the above proposition) so that all the finite dimensional distributions P_n of $x_1, x_2, \ldots x_n$ are well defined. It remains to prove that $P\{x \,|\, \sum_1^\infty |x_\nu| < \infty\} = 1$. Introduce the function

$$q(n, \varepsilon) = \int_m \varphi(t) \lambda(dt)$$

$$= \int_{x \in R^n} \int_{t \in R^n} \exp \left[i \sum_1^n t_\nu x_\nu \right] P_n(dx) \prod_1^n \frac{1}{\pi \varepsilon_n \left[1 + \dfrac{t_\nu^2}{\varepsilon_n^2} \right]} dt_1 \ldots dt_n$$

$$= \int_{x \in R^n} e^{-\varepsilon_n \sum_1^n |x_\nu|} P_n(dx) = E \, e^{-\varepsilon_n \sum_1^n |x_\nu|}.$$

Then λ-continuity implies that

$$E e^{-\varepsilon_n \sum_1^n |x_\nu|} = q(n, \varepsilon_n) \to 1, n \to \infty,$$

so that
$$\varepsilon_n \sum_1^n |x_\nu| \to 0 \text{ in probability, } n \to \infty.$$

But this implies that $S_n = \sum_1^n |x_\nu|$ converges (to a finite limit) in probability, since otherwise we could find, for any $\delta > 0$, a sequence $A_n \uparrow + \infty$ such that $P(S_n > A_n) > \delta$. Choosing $\varepsilon_n = 1/A_n \downarrow 0$ this means that $P(\varepsilon_n S_n > 1) > \delta$ for large n which is impossible. This shows that $P\{x | \sum_1^\infty |x_\nu| < \infty\} = 1$.

Now suppose instead that P is a well-defined probability measure in l_1. Then

$$1 \geqslant \int \varphi(t) \lambda(dt) = E e^{-\sum_1^n \varepsilon_\nu |x_\nu|} \geqslant E e^{-\max \varepsilon_\nu \|x\|} \to 1$$

as $n \to \infty$, since $\|x\|$ is an ordinary stochastic variable. Hence $\varphi(t)$ is λ-continuous.

This reasoning is of general scope. In the Hilbert space situation we could do the analogous thing, using as Λ a set of independent and normal distributions $N(0,\sigma)$ for the t_ν's with $\sum_1^\infty \sigma_\nu^2 < \infty$. This would lead us back to the S-topology. In a general context we would embed the space X in a bigger space X', perhaps given in coordinate form. We would then try to approximate $I_X(x')$, the indicator function of X in X', by more regular functions $I_X^{(n)}$ and express $E I_X(x') = P(X)$ through the Fourier transform of $I_X^{(n)}$ and $\varphi(x^*)$.

Considerations of this type are of some interest for clarifying the logical difficulties, but the resulting criteria do not seem to be of much practical use because of their indirect form.

Let us now recall the comments made at the end of section 4.2. and see how they can be adjusted to hold in the case of Hilbert spaces. Assume that our Hilbert space X is represented as a direct sum

$$X = X_1 \oplus X_2 \oplus X_3 \oplus \ldots$$

of elements $x = (x_1, x_2, \ldots)$ with x_k in the Hilbert spaces X_k with the inner product $(a,b)_k$ and norm $\|a\|_k = \sqrt{(a,a)_k}$. In X we have the inner product

$$(x, y) = \sum_{k=1}^\infty (x_k, y_k)_k.$$

Suppose that on the subspaces $Y_n = X_1 \oplus X_2 \oplus \ldots \oplus X_n$ we have some probability structure and we want to induce one in X. Put $\|y_n\|_n^2 = \sum_{\nu=1}^n \|x_\nu\|_\nu^2$.

Theorem 6.2.5. Suppose that a sequence of probability measures $P_1^{(n)}$, $P_2^{(n)}, \ldots$ is defined consistently on each Y_n, $n = 1, 2, \ldots$ Introduce the bounded non-decreasing functions

$$F_k(u) = \lim_{n \to \infty} P_k^{(n)} \{ y_n \,|\, \|y_n\|_n \leqslant u \}.$$

If $F_k(+\infty) = \lim_{u \to +\infty} F_k(u) = 1$ then the $P_k^{(n)}$ define probability distributions P_k in X. If $F_k(u) \to 1$ for any $u > 0$ when $k \to \infty$ then $P_k \to \delta_0$.

If $P_t^{(n)}$ are consistently defined homogeneous processes in Y_n and

$$F_t(u) = \lim_{n \to \infty} P_t^{(n)} \{ y_n \,|\, \|y_n\|_n \leqslant u \},$$

satisfies $F_t(+\infty) = 1$ and $F_t(u) \to 1$ for every $u > 0$ when $t \downarrow 0$, then $P_t^{(n)}$ define a continuous, homogeneous process P_t in X.

Proof: We have assumed the measures $P_k^{(n)}$ to be consistent so that $P_k^{(n')}$ is the projection of $P_k^{(n'')}$ on $Y_{n'}$, if $n' < n''$. Then we have a probability measure on the space of sequences $(x_1, x_2, \ldots)$. Consider the functions

$$P_k^{(n+1)} \{ y_n \,|\, \|y_n\|_{n+1} \leqslant u \} = P_k^{(n+1)} \{ y_n \,|\, \|y_n\|_n^2 + \|x_{n+1}\|^2 \leqslant u \} \leqslant P_k^{(n)} \{ y_n \,|\, \|y_n\|_n \leqslant u \}$$

which form a non-increasing sequence when n increases. Hence $F_k(u)$ is well defined. We just have to show that

$$P_k \left\{ x \,\Big|\, \sum_1^\infty \|x_k\|_k^2 < \infty \right\} = 1$$

which follows from

$$P_k \left\{ x \,\Big|\, \sum_1^\infty \|x_k\|_k^2 < \infty \right\} = \lim_{n \to +\infty} P_k \left\{ x \,\Big|\, \sum_1^\infty \|x_k\|_k^2 < n \right\} = \lim_{n \to +\infty} F_k(n) = 1.$$

Further

$$P_k \{ x \,|\, \|x\| < \varepsilon \} = F_k(\varepsilon) \to 1$$

for any $\varepsilon > 0$ when $k \to \infty$ so that $P_k \to \delta_0$.

Applying this to a $P_t^{(n)}$ the rest of the theorem follows directly.

Remark. It is sufficient (but not necessary) for the first part of the theorem to assume that

$$\sigma_k^2 = \lim_{n \to \infty} \int_{Y_n} \|y_n\|_n^2 \, P_k^{(n)} (dy_n)$$

is finite and $\sigma_k^2 \to 0$ as $k \to \infty$. Similarly for the second part.

An interesting application is to $X = l_2$, where we choose the X_n as the one-dimensional spaces corresponding to the coordinates of l_2. In each X_n we know the form of a continuous, homogeneous process (or rather its infinitesimal generator) and these can then be used as building blocks to construct a P_t in $X = l_2$.

140   *Stochastic linear spaces*

6.3. Normal distributions in a Hilbert space

Now it is not difficult to see what the correct definition should be of a *normal probability distribution* in the Hilbert space X.

Definition 6.3.1. By a normal distribution P in a Hilbert space X we mean the distribution uniquely determined by the Fourier transform

$$\hat{P}(x^*) = \exp\left\{i(x^*, m) - \tfrac{1}{2}(Sx^*, x^*)\right\}.$$

Here m is a fixed element in X and S is an S-operator.

Theorem 6.2.4 shows that there is such a distribution and the uniqueness is clear from Theorem 6.2.1, statement 6). It can also be seen that the restriction on the operator S appearing in the quadratic form in the exponent is necessary.

We now have the straight forward identification of m and S in terms of the moments of the distribution.

Theorem 6.3.1. For the normal distribution m is the mean value and S is the covariance operator.

Theorem 6.3.2. If x is a stochastic element in a Hilbert space X with a normal distribution, then x can be written as

$$x = m + \sum_{\nu} \xi_{\nu} e_{\nu},$$

where the e_{ν} are orthogonal unit vectors in X and the ξ_{ν} are independent stochastic variables distributed $N(0, \sigma_{\nu})$, where σ_{ν}^2 are the eigenvalues of the operator S. The sum is finite or denumerable and converges (strongly) with probability one.

Proof of Theorem 6.3.2: Choose orthonormal vectors $\{e_{\nu}\}$ so that S has diagonal form in the corresponding coordinate system. Since S has finite trace, its eigenvalues σ_{ν}^2 satisfy $\sum_{\nu} \sigma_{\nu}^2 < \infty$. Choose the ξ's as described and form the sum $y = \sum_{\nu} \xi_{\nu} e_{\nu}$ which converges in the mean with probability one. Using Theorem 6.2.1, statements 3 and 7, it follows that the Fourier transform of the distribution of y is

$$\prod_{\nu} \exp\left[-\frac{\sigma_{\nu}^2}{2}(x_{\nu}^*)^2\right] = \exp\left[-\frac{1}{2}\sum_{\nu}\sigma_{\nu}^2(x_{\nu}^*)^2\right] = \exp\left[-\frac{1}{2}(Sx^*, x^*)\right]$$

and similarly for $x = m + y$.

Proof of Theorem 6.3.1: Use the representation of x. We then get

$$\begin{cases} Ex = m \\ E(x^*, x - m)^2 = \sum_\nu E(x_\nu^*, \xi_\nu)^2 = \sum_\nu \sigma_\nu^2 (x_\nu^*)^2 = (Sx^*, x^*) \end{cases}$$

as stated; in both relations we have to change the order between the expectation and summation symbol but the justification of this is straightforward.

Now we just go ahead.

Theorem 6.3.3. If x and y are two normal, independent, stochastic elements, in a Hilbert space, with the respective mean values m_x and m_y and covariance operators S_x and S_y, then $x+y$ has also a normal distribution with the mean value $m_x + m_y$ and the covariance operator $S_x + S_y$.

The proof is immediate.

Theorem 6.3.4. (*à la Cramér*). If $z = x + y$, x and y independent stochastic elements in X, and z normal, then x and y must have a normal distribution.

Proof: It is not difficult to see that both x and y must have existing mean values and covariance operators; let us consider them as given. For an arbitrary $x^* \in X$, the scalar stochastic variables $x^*(x)$ and $x^*(y)$ are known to be normal, and since their moments of order one and two can be computed from the fixed mean values and covariance operators their complete distributions are determined. But knowing these distributions we can construct the entire probability distributions of x and y as we have observed before and this completes the proof (see Notes 6.3.$_1$).

Theorem 6.3.5. If B is a bounded linear transformation from X to X and x is a stochastic element with a normal distribution with mean value m and covariance operator S then $y = Bx$ has a normal distribution with the parameters

$$\left. \begin{array}{l} m_y = Bm \\ S_y = BSB^* \end{array} \right\}.$$

Proof: For any x^* the stochastic variable $(x^*, y) = (x^*, Bx) = (B^* x^*, x)$. The element $y = Bx$ has a well defined probability distribution. Further its Fourier transform is given by

$$E \exp i(x^*, y) = \exp i(B^* x^*, x) = \exp \left\{ i(B^* x^*, m) - \tfrac{1}{2}(SB^* x^*, B^* x^*) \right\}$$
$$= \exp \left\{ i(x^*, Bm) - \tfrac{1}{2}(BSB^* x^*, x^*) \right\},$$

which proves the statement.—It could also have been shown directly that $T = BSB^*$ is an S-operator. Indeed, it is immediate that T is Hermitian, bounded and non-negative definite. It is completely continuous, since it transforms a weakly convergent sequence into a strongly convergent one. To show that T has finite trace it is easiest to use coordinate language, choose a coordinate system diagonalizing S and compute the trace.

Theorem 6.3.6. Let $P_n, n = 1, 2, \ldots$, be normal distributions in a Hilbert space X with mean values m_n and covariance operators S_n. If $m_n \to m$ (strongly), $(S_n x^*, x^*) \to (S x^*, x^*)$ for every $x^* \in X$ and $S_n \leqslant T$, where T is some S-operator, then P_n converges weakly to a normal distribution with mean value m and covariance operator S.

Proof: We can treat m_n and m as if they vanished, this means just a translation (or rather a strongly convergent sequence of translations) in the Hilbert space. For any $x^* \in X^*$

$$\hat{P}_n(x^*) = \exp\left\{ -\tfrac{1}{2}(S_n x^*, x^*) \right\} \to \exp\left\{ -\tfrac{1}{2}(S x^*, x^*) \right\} = \hat{P}(x^*).$$

Now use Theorem 6.2.3. We have

$$E_n x_\nu^2 = \int_X x_\nu^2 P_n(dx) = s_\nu^n = (\nu\nu)\text{th element of } S_n \text{ so that}$$

$$E_n \sum_{N+1}^{\infty} x_\nu^2 = \sum_{N+1}^{\infty} s_\nu^n \leqslant \sum_{N+1}^{\infty} t_\nu,$$

where t_ν is the $(\nu\nu)$th element of T. Hence

$$\lim_{N\to\infty} \sup_n E_n \sum_{N+1}^{\infty} x_\nu^2 \leqslant \lim_{N\to\infty} \sum_{N+1}^{\infty} t_\nu = 0$$

and condition 1) of the theorem mentioned is satisfied, so that we have proved the result (see Notes 6.3.$_2$).

In this section we have studied normal distributions in a Hilbert space only. In a general Banach space one could, in analogy with the Hilbert space situation, define a normal stochastic element x as one for which $x^*(x)$ is a normal stochastic variable for every $x^* \in X^*$. Some of what has been said above would still apply, but at present our knowledge is incomplete as regards normal distributions in general Banach spaces.

6.4. The law of large numbers

To start this section let us make some elementary considerations. If $x_1, x_2, x_3, \ldots$ are independent stochastic elements in a Banach space X with the same probability distribution and the mean value m, let us form the average

$$\bar{x} = \frac{1}{n}(x_1 + x_2 + \ldots + x_n).$$

For any $x^* \in X^*$ we hawe

$$x^*(\bar{x}) = \frac{1}{n}[x^*(x_1) + x^*(x_2) + \ldots + x^*(x_n)],$$

and the scalar stochastic variables $x^*(x_\nu)$ are also independent and have the same distribution. Their mean value exists and equals $x^*(m)$. It then follows from the strong law of large numbers that the stochastic variables $x^*(\bar{x})$ converge to their mean value $x^*(m)$ with probability one. Using the assumed separability of X^* we can choose a denumerable sequence $x_1^*, x_2^*, \ldots$ dense in X^*. By a standard argument it then follows that almost certainly $\bar{x}$ converges weakly to m.

But this is not a really essential extension of the classical law of large numbers, since it just concerns the (simultaneous) asymptotic properties of $\bar{x}$ measured in the directions of the various x^*. To extend this to more stringent modes of convergence we first deal with a simple result that is based upon

Lemma 6.4.1. (*Tchebychev's inequality*). Let x be a stochastic element in a Hilbert space X with $E\|x\|^2 < \infty$ and $Ex = 0$ (the second condition just means that we have chosen Ex as the origin by a translation of X). Then

$$P\{\|x\| \geq C\} \leq \frac{E\|x\|^2}{C^2}.$$

Proof as in the scalar case.

Theorem 6.4.1. (*Weak law with strong convergence.*) Let $x_1, x_2, \ldots, x_n$ be uncorrelated stochastic elements in a Hilbert space with common $E\|x\|^2 < \infty$ and denote $Ex = m$. Then for any $\varepsilon > 0$

$$P\{\|\bar{x} - m\| > \varepsilon\} \to 0 \text{ as } n \to \infty.$$

Proof: That x_ν and x_μ are uncorrelated should be understood in the sense that $E(x_\nu, x_\mu) = (Ex_\nu, Ex_\mu)$.

The proof is completely straightforward. We have $E\|\bar{x}-m\|^2 = (1/n)\,E\|x-m\|^2$ so that the lemma gives us

$$P\{\|\bar{x}-m\|\geqslant\varepsilon\}\leqslant\frac{E\|x-m\|^2}{n\,\varepsilon^2}\to 0$$

as stated.

If $x_1, x_2, \ldots$ were independent and identically distributed we could just as well have used the Fourier transform of the distribution of $\bar{x}$ which is

$$\hat{P}(x^*)=\left[1+\frac{i\,x^*(m)}{n}+O\left(\frac{1}{n^2}\right)\right]^n\to e^{i\,(x*,\,m)}.$$

This convergence guarantees the result via Theorem 6.2.3., since in this case

$$R_N\leqslant\sum_N^\infty m_\nu^2+\sum_N^\infty E(\xi_\nu-m_\nu)^2\downarrow 0 \text{ as } N\to\infty.$$

Here we have written m and x in coordinates $m=(m_1,m_2,\ldots)$ and $x=(\xi_1,\xi_2,\ldots)$.

The following result is more essential.

Theorem 6.4.2. (*Strong law with strong convergence.*) Let $x_1,x_2,x_3,\ldots$ be independent, identically distributed stochastic elements in a Banach space X and with $E\|x\|<\infty$. Then $\bar{x}$ converges strongly with probability one to the mean value Ex.

Proof: Let us assume to begin with that the stochastic elements can take only denumerably many values $\eta_1,\eta_2,\eta_3,\ldots$ Choose an integer k and define

$$x_\nu^k=\begin{cases} x_\nu \text{ if } x_\nu=\eta_1,\eta_2,\ldots \text{ or } \eta_k \\ 0 \text{ otherwise}\end{cases}$$

and $x_\nu=x_\nu^k+r_\nu^k$. We have

$$\bar{x}=\frac{1}{n}\sum_{\nu=1}^n x_\nu^k+R_n^k,\ R_n^k=\frac{1}{n}\sum_{\nu=1}^n r_\nu^k.$$

The first sum converges almost certainly (strongly) for any k to Ex^k. Actually this is just a finite dimensional version of the classical strong law of large numbers. Also with probability one

$$\frac{1}{n}\sum_{\nu=1}^n \|r_\nu^k\|\to E\|r_\nu^k\|, \text{ which does not depend upon } \nu.$$

But $E\|r_\nu^k\|\to 0$ as $k\to\infty$. Combining these statements we see that $\bar{x}$ converges almost certainly to Ex.

Now the general case. Using the separability of X we choose a denumerable sequence of points $x_1, x_2, \ldots$ everywhere dense in X. Each x_i we circumscribe spheres $S_i(\varepsilon) = \{x \mid \|x - x_i\| \leqslant \varepsilon\}$ so that X is covered by a system of overlapping spheres. Introduce the disjoint sets

$$\left\{ \begin{aligned} &E_1(\varepsilon) = S_1(\varepsilon) \\ &E_2(\varepsilon) = S_2(\varepsilon) \cap S_1^*(\varepsilon) \\ &E_3(\varepsilon) = S_3(\varepsilon) \cap S_1^*(\varepsilon) \cap S_2^*(\varepsilon). \\ &\ldots \end{aligned} \right.$$

Define the transformation $x \to T_\varepsilon x$, $T_\varepsilon x = x_i$, where x_i is the center of $S_i(\varepsilon)$ if $x \in E_i(\varepsilon)$. It is clear that the stochastic elements $T_\varepsilon x_\nu$ have the same properties as the x_ν in the first part of this proof. Hence for any $\varepsilon > 0$

$$\frac{1}{n} \sum_{\nu=1}^{n} x_\nu = \frac{1}{n} \sum_{\nu=1}^{n} T_\varepsilon x_\nu + \frac{1}{n} \sum_{\nu=1}^{n} (T_\varepsilon x_\nu - x_\nu);$$

The first term on the right side converges almost certainly to $ET_\varepsilon x$ and the norm of the second term is at most ε. We now choose a sequence $\varepsilon_1, \varepsilon_2, \ldots \downarrow 0$ and the proof is complete.

Other modes of convergence may be of interest, such as the following one.

Theorem 6.4.3. (*Convergence in the mean, order α.*) Let $x_1, x_2, \ldots$ be a sequence of independent and identically distributed stochastic elements in the Banach space X. If $E\|x\|^\alpha < \infty$, $1 \leqslant \alpha < \infty$, then

$$\lim_{n \to \infty} E\|\bar{x} - m\|^\alpha = 0.$$

6.5. The central limit theorem

The linear structure of the underlying space X makes it possible to formulate and prove analogs of the central limit theorem. The theory is far from complete at present, but certain partial results exist and have considerable interest in spite of their limitation. The first one parallels the classical case closely.

Theorem 6.5.1. Consider a sequence $x_1, x_2, x_3, \ldots$ of independent and identically distributed stochastic elements in a Hilbert space X. Suppose that $E\|x\|^2 < \infty$ and introduce the mean value $Ex = m$ and the covariance operator S. Then the normed stochastic element

$$y_n = \frac{1}{\sqrt{n}} \sum_{\nu=1}^{n} (x_\nu - m)$$

has a distribution P_n that converges (weakly) to the normal distribution in X with mean value zero and the covariance operator S.

Proof: We use the Fourier transform $\hat{P}_n(x^*)$. For a given x^* this is practically the same thing as the ordinary characteristic function of the stochastic variable

$$(x^*, y_n) = \frac{1}{\sqrt{n}} \sum_{\nu=1}^{n} (x^*, x_\nu - m).$$

But we know from the classical case that the limit of the characteristic function of these stochastic variables is of the form

$$\hat{P}_n(x^*) \to e^{-\frac{1}{2} \text{ variance of } (x^*,\, x_\nu - m)} = e^{-\frac{1}{2}(Sx^*,\, x^*)}.$$

To apply Theorem 6.2.3. we have to verify condition 1) in that theorem. Writing $y_n = (y_n^1, y_n^2, \ldots)$ we have

$$E \sum_{\nu=N}^{\infty} (y_n^\nu)^2 = \sum_{\nu=N}^{\infty} S_{\nu\nu}$$

(the covariance operator must be an S-operator) so that

$$\lim_{N \to \infty} R_N = \lim_{N \to \infty} \sum_{\nu=N}^{\infty} S_{\nu\nu} = 0$$

which proves the statement made.

It is also possible to prove the following (see Notes 6.5).

Theorem 6.5.2. Let $x_1, x_2, x_3, \ldots$ be independent and identically distributed stochastic elements in a reflexive G-space with a basis. Assume that $E\|x\|^2 < \infty$ and denote Ex by m. The probability distribution of the normed variable y_n (defined as in the previous theorem) then converges (weakly) to a normal distribution.

6.6. Stochastic Schwartz distributions

Let Φ be the set of all infinitely differentiable functions on R^1 vanishing outside of compact sets and denote by D the set of *Schwartz distributions* on R^1. We assume that the basic facts about such distributions are known to the reader.

To introduce and to operate with probability measures on D we cannot just apply the theory discussed in this chapter. It is true that D has a linear structure but it is *not* a Banach space. The necessary modifications are of some interest and we shall discuss this briefly.

Denote by $\mathcal{D}$ the σ-algebra generated by sets of the form $\{d \,|\, d(\varphi) \leqslant c\} \subset D$ for arbitrary $\varphi \in \Phi$ and $c \in R^1$.

Definition 6.6.1. By a stochastic Schwartz distribution we mean a variable taking values in D according to a probability measure defined in $\mathcal{D}$.

An immediate consequence of this definition is that if d_1 and d_2 are two stochastic Schwartz distributions then this is also true about an arbitrary linear combination $c_1 d_1 + c_2 d_2$. Also if $d_n, n = 1, 2, \ldots$, is a sequence of stochastic Schwartz distributions converging almost certainly, (in the Schwartz topology) then the limit is also a stochastic Schwartz distribution. Any derivative $D^n d$ of a stochastic Schwartz distribution is a stochastic Schwartz distribution, which can be seen from the relation $D^n d(\varphi) = (-1^n)(D^n \varphi)$.

Definition 6.6.2. A stochastic Schwartz distribution d is said to have the mean value $m, m \in D$ if for every $\varphi \in \Phi$ we have $Ed(\varphi) = m(\varphi)$.

Obviously if d_1 and d_2 have the mean values m_1 and m_2 then a linear combination $c_1 d_1 + c_2 d_2$ has the mean value $c_1 m_1 + c_2 m_2$. Also if $Ed = m$, then the nth derivative d_n of d has the mean value $D^n m$; indeed

$$(-1)^n D^n m(\varphi) = m(D^n \varphi) = Ed(D^n \varphi) = (-1)^n E D^n d(\varphi)$$

so that $E D^n d = D^n Ed = D^n m$.

A law of large numbers can be stated as follows. Consider a sequence of independent (independence is defined in the obvious manner) and identically distributed stochastic Schwartz distributions $d_1, d_2, d_3, \ldots$ with the mean value $Ed_n = m$. We demand that for every realization the sequence is equicontinuous at the origin: for any sequence $\varphi_i \in \Phi$ converging to zero, $\varphi_i \to 0$ as $i \to \infty$, and for any positive ε we have for sufficiently large values of i

$$|d_n(\varphi_i)| \leqslant \varepsilon \text{ all } n.$$

We can then state the theorem.

Theorem 6.6.1. If d_n are independent, identically distributed, with mean value m and such that all realizations are equicontinuous at the origin, then

$$\lim_{N\to\infty} \frac{1}{N} \sum_{n=1}^{N} d_n = m$$

with probability one.

For a proof see Notes 6.6.$_1$.

This theory is still in a preliminary state.

6.7. Illustrations

It is easy to exemplify the notion of expected values in a Banach space. Take e.g. $X = c =$ the set of all convergent sequences $x = (x_1, x_2, ...)$ of real numbers with the norm of uniform convergence. Suppose that

$$E\|x\| = E \sup_n |x_n| < \infty.$$

Then the mean value of the stochastic element x is simply the constant element m in X with coordinates $m = (m_1, m_2, ...)$, $m_n = Ex_n$. Since almost certainly $\lim_n x_n$ exists and since the stochastic variable $\sup_n x_n$ has finite expectation it follows (from dominated convergence) that $\lim_n m_n$ exists so that $m \in c$. This element of c must be the expected value of x, since the dual space $c^* = l_1$ consists of summable sequences $y = (y_0, y_1, y_2, ...)$, $\sum_0^\infty |y_n| < \infty$. Hence for any $y \in c^*$ we have

$$y(m) = y_0 \lim m_n + \sum_1^\infty y_n m_n = E[y_0 \lim x_n + \sum_1^\infty y_n x_n] = Ey(x)$$

which verifies the statement.

Or take instead $X = l_2$. If

$$E\|x\| = E\sqrt{\sum_1^\infty x_n^2} < \infty$$

then the vector $m = (m_1, m_2, ...)$ with $m_n = Ex_n$ belongs to l_2; indeed for any N we have

$$\sum_1^N m_n^2 = E \sum_1^N m_n x_n \leqslant \left(\sum_1^N m_n^2\right)^{1/2} E\|x\|$$

so that $\sum_1^\infty m_n^2 < \infty$. But for any $y = (y_1, y_2, ...) \in l_2^* = l_2$

$$y(m) = \sum_1^\infty y_n m_n = \sum_1^\infty y_n Ex_n = E\sum_1^\infty y_n x_n = Ey(x),$$

where the next last equality follows from dominated convergence and Schwarz' inequality. Hence $Ex = m$.

Finally take $X = C(0,1) =$ the set of continuous real-valued functions on the interval $(0,1)$ with $\|x\| = \max\limits_{0 \leqslant t \leqslant 1} |x(t)|$, $x = x(t)$. If

$$E\|x\| = E \max_{0 \leqslant t \leqslant 1} |x(t)| < \infty$$

then the function $m = m(t) = Ex(t)$ belongs to $X = C(0,1)$ since

$$\lim_{h \to 0} |m(t+h) - m(t)| \leqslant \lim_{h \to 0} E|x(t+h) - x(t)| = 0$$

again because of dominated convergence. Since C^* consist of bounded measures μ on $(0,1)$ we have for any $\mu \in C^*$

$$\mu(m) = \int_0^1 m(t)\,\mu(dt) = \int_0^1 Ex(t)\,\mu(dt) = E \int_0^1 x(t)\,\mu(dt) = E\mu(x)$$

so that $Ex = m$.

In this connection it may be remarked that, since any separable Banach space is isometric and isomorphic to some sub-space of $C(0,1)$ (this is the well-known theorem of Banach and Mazur, see Banach [1]), the probability theory on a separable Banach space deals with problems concerning strictly continuous stochastic processes and is therefore of special interest.

The most direct and perhaps also the most important of the applications of this chapter is to the theory of *stochastic processes*. A stochastic process is usually defined as a function $x(t,\omega)$ of two arguments, time t and the stochastic parameter ω, with certain measurability assumptions. In the linear theory of stochastic processes it is customary to consider $x(t,\omega)$ as a set of measurable functions (stochastic variables) of ω, $x_t(\omega)$, one for each value of t. If these functions of ω can be identified with elements of some Banach space the process forms a curve with the parameter t in this Banach space. Of course, one can also go the opposite way. For each ω we get a function $x_\omega(t)$ of the real parameter t and now we let these functions belong to some Banach space. Then the process consists of a probability distribution defined over a Banach space and this was the starting point of the present chapter.

Let $x_\omega(t)$ be a stochastic process defined via a probability distribution over the Hilbert space $L_2(0,1)$ with

$$x_\omega(t) = m(t) + y_\omega(t).$$

Here $m(t)$ is a fixed element in $L_2(0,1)$, $y_\omega = y_\omega(t)$ has $E\|y_\omega\|^2 < \infty$ and $Ey_\omega = 0$. Now take repeated, independent observations on $x_\omega(t)$, denoting the sample functions by $x^1(t), x^2(t), \ldots x^n(t)$. Form the average

$$m^* = m^*(t) = \frac{1}{n} \sum_{\nu=1}^{n} x^\nu(t),$$

which is also an element in the Hilbert space. We then know from Theorem 6.4.2. that, if we exclude an event of probability zero, the estimate m^* converges strongly to m as n tends to infinity. In other words, the function $m(t)$ is estimated consistently (in the strong sense used) by the average $m^*(t)$. Using Theorem 6.5.1 we can study the convergence of the distributions of m^* by means of the central limit approximation. It is evident that this approach is useful in statistical studies of stochastic processes.

Now let us look at something very similar. If $F(x)$ is a continuous distribution function of a real stochastic variable x we make the usual transformation $y = F(x)$ to a rectangular distribution. Having observed a sample of n values $y_1, y_2, ..., y_n$ we form the empirical distribution function

$$F^*(u) = \frac{\text{number of } y_\nu\text{'s} \leqslant u}{n} = \frac{1}{n} \sum_{\nu=1}^{n} \varepsilon_{y_\nu}(u),$$

where $\varepsilon_y(u) = 1$ if $y \leqslant u$ and 0 otherwise. The $\varepsilon_y(u)$ are stochastic elements in $L_2(0, 1)$ with

$$\begin{cases} E\,\varepsilon(u) = u \\ E\,\|\varepsilon(u)\|^2 = \tfrac{1}{2} < \infty. \end{cases}$$

Thus the element $F^*(u) \in L_2(0,1)$ converges almost certainly in the norm to the true distribution function u. We can also deduce from Theorem 6.5.1 that if $f(z)$ is a real valued continuous functional defined in $L_2(0,1)$, then the distribution of the stochastic variable

$$f_n = f\left(\frac{F^*(u) - u}{\sqrt{n}}\right)$$

converges to that of $f(Z)$, where Z is a stochastic element in the Hilbert space with a normal distribution. The mean value of Z is zero and its covariance operator S is given by

$$(Sz^*, z^*) = \int_0^1 \int_0^1 z^*(u)\,z^*(v)\,K(u, v)\,du\,dv,$$

where $z^* = z^*(u) \in L_2(0,1)$ and $K(u,v) = \min(u,v) - uv$.

The convergence of the empirical distribution functions can be studied more generally. We can use other topologies in the convergence statement. We can also try to extend it to more general spaces than the real line. Certain results of this character have been obtained, see Notes 6.7. Let S be a complete and separable space with a metric $d(s', s'')$. A probability measure P is defined on the Borel sets of S and we demand that $\int_s d(s_0, s) P(ds) < \infty$. Taking n independent observations from the P-population, $s_1, s_2, ..., s_n$, we form the probability measure

$$P^* = \frac{1}{n} \sum_{\nu=1}^{n} \delta_{s_\nu}.$$

Just as above we can say that P^* approximates P if n is large enough, and we can make this into a statement about consistent estimation as follows. Let F be the set of real functions $f(s)$ on S such that

$$\sup \frac{|f(s') - f(s'')|}{d(s', s'')} < \infty.$$

With the obvious norm, F is a Banach space. Now it can be seen that set functions of the type $P^* - P$ belong to a certain subspace Φ of the Banach space F^*, and we can now apply the various laws of large numbers of section 6.4. It follows e.g. that (almost certainly) $\|P^* - P\| \to 0$ as $n \to \infty$ in the norm of Φ. In the special case $S = R^1$ this takes the attractive form that

$$\lim_{n \to \infty} \int_{-\infty}^{\infty} |F^*(y) - F(y)| \, dy = 0$$

under the condition

$$\int_{-\infty}^{\infty} |y| \, dF(y) < \infty.$$

CHAPTER 7

—

STOCHASTIC ALGEBRAS

7.1. Additive and multiplicative limit theorems

In the last chapter we let our stochastic elements take values in a (separable) Banach space X and used the linear properties of X to formulate and prove a number of results. We now assume that X forms a (separable) *topological algebra*. We have then also access to multiplication xy, x and $y \in X$, a continuous binary operation, not always commutative. Let x and y be independent stochastic elements in X; we can then form the new stochastic element $z = xy$. Denoting the respective probability distributions by P_x, P_y and P_z, we see (see Notes 7.1.$_1$) that P_z is a well defined probability measure. We shall write $P_z = P_x \circ P_y$. It is clear that we can study problems expressed in the $\circ$-operation just as we have studied the similar problems involving the $*$-operation.

What is perhaps more essential is to find out what logical relations hold between these two types of results. To be specific let us mention one particularly important problem. If $P_1^{(n)}, P_2^{(n)}, ..., P_n^{(n)}$ are probability measures belonging to $\mathcal{P}(X)$ we can form the two new probability distributions

$$Q_n = P_1^{(n)} * P_2^{(n)} * ... * P_n^{(n)}$$
$$R_n = P_1^{(n)} \circ P_2^{(n)} \circ ... \circ P_n^{(n)}$$

If we can state limit laws for one of Q_n or R_n, what can we infer about the other one? Especially Q_n can often be dealt with successfully since addition is commutative in the algebra, and this leads us back to problems in a Banach space like those studied in Chapter 6. In this section we shall prove some useful relations of this type.

First some terminology. Consider a stochastic process $x(t)$, $0 \leqslant t < \infty$, taking values in X and with the probability distributions $P_t \in \mathcal{P}(X)$. The process should have independent "increments" (in the additive or multiplicative sense), and the probability distribution of an increment should only depend upon the length of the corresponding time interval. If we have independent increments so that $P_s * P_t = P_{s+t}$; $s, t \geqslant 0$; we shall call

$\{P_t\}$, or $x(t)$ a *homogeneous additive process*. If instead we have independent multiplicative increments so that $P_s \circ P_t = P_{s+t}$; $s, t \geqslant 0$; we shall call $\{P_t\}$, or $x(t)$ a *homogeneous multiplicative process*. In addition we will also impose some continuity property for the process.

Let us first assume that X is a *Banach algebra*, so that its norm satisfies $\|xy\| \leqslant \|x\| \cdot \|y\|$, and has a unit element e. Let $y(t)$ be a homogeneous additive process taking values in X and satisfying the continuity condition

$$C_1 : \sum_\nu E\|y(t_\nu) - y(t_{\nu-1})\| \leqslant M < \infty$$

for every division $t_0 = 0, t_1, t_2, ..., t_{n-1}, t = t_n$ of the fixed interval $(0,t)$. We shall now construct a homogeneous multiplicative process $x(t)$ related in a natural way to $y(t)$. Write down the formal expression (see Notes 7.1.$_2$)

$$x(t) = e + \int_0^t dy(s) + \iint\limits_{0<s_1<s_2<t} dy(s_1)\,dy(s_2)$$

$$+ \iiint\limits_{0<s_1<s_2<s_3<t} dy(s_1)\,dy(s_2)\,dy(s_3) + ...$$

which we now shall give a precise meaning.

The integrals shall be understood as limits in the strong $L_1(X)$ topology of ordinary Riemann-Stieltjes sums. We shall discuss the second (the double) integral which is typical. For a division with the points $\{t_\nu\}$ write down the Riemann-Stieltjes sum

$$S = \sum_{\nu<\mu} [y(t_\nu) - y(t_{\nu-1})]\,[y(t_\mu) - y(\mu-1)],$$

where the order of the two factors may be essential. As the division is made finer these sums will converge to a limit which does not depend upon what sequence of division we have chosen. To prove this let us take another division $t_0' = 0 < t_1' < t_2' < ... < t_{m-1}' < t = t_m'$ and denote the corresponding sum by S'. The combined division $t_0'' = 0 < t_1'' < t_2'' < ... < t$ gives rise to a sum S''. Consider now the figure below.

To each rectangle of the figure corresponds one term in the respective sum. In the difference $S'' - S$ only the shaded rectangles in the upper part of the figure occur which follows from $(a+b)(c+d) = ac + bc + bd + ad$. It is now clear what happens in general: the difference $S'' - S$ consists of rectangles close to the diagonal and its norm is dominated by

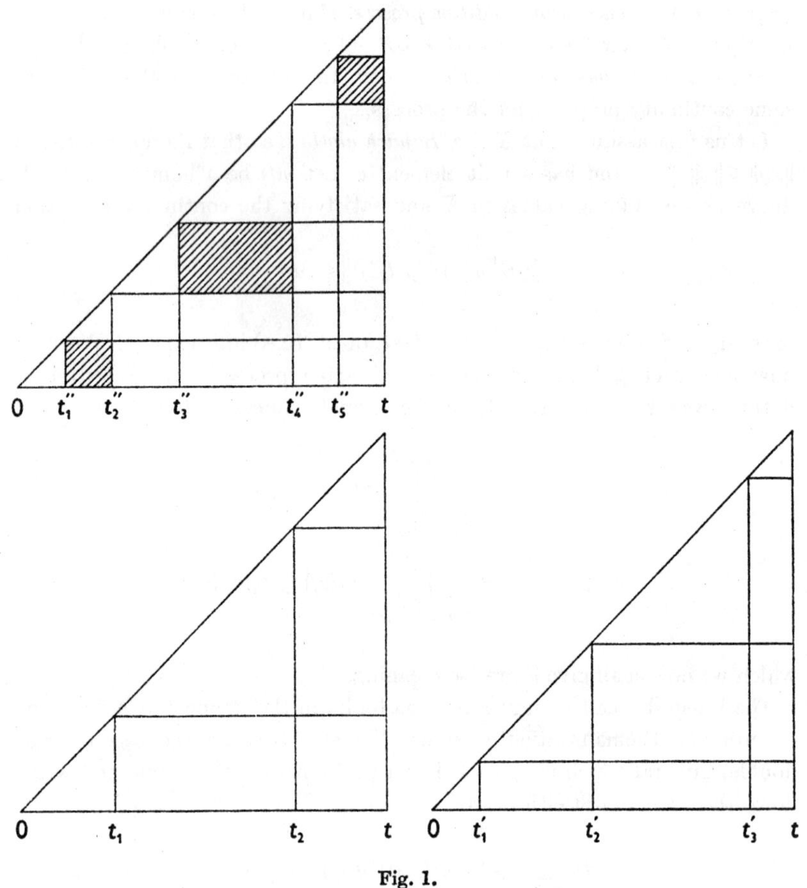

Fig. 1.

$$\|S'' - S\| \leqslant \sum_{\nu} \left\| [y(t_\nu'') - y(t_{\nu-1}'')]\,[y(t_{\nu+1}'') - y(t_\nu'')] \right\|$$

$$\leqslant \sum_{\nu} \|y(t_\nu'') - y(t_{\nu-1}'')\| \cdot \|y(t_{\nu+1}'') - y(t_\nu'')\|$$

so that the $L_1(X)$-norm of the difference is

$$E\,\|S'' - S\| \leqslant \sum_{\nu} E\,\|y(t_\nu'') - y(t_{\nu-1}'')\| \cdot E\,\|y(t_{\nu+1}'') - y(t_\nu'')\|$$

which tends to zero with $\max_{\nu}(t_{\nu+1}'' - t_\nu'')$. Thus S'' differs arbitrarily little from S and hence also from S' which shows that the double integral is uniquely defined. But we also get a bound for the $L_1(X)$-norm of the integrals by just looking at the Riemann sums

$$E\left\|\ \underset{0<s_1<s_2<...<s_k<t}{\int\!\!\int\cdots\int}\ dy(s_1)\,dy(s_2)\ldots dy(s_k)\right\|\leqslant\frac{M^k}{k!}.$$

Hence the sum defining $x(t)$ converges in the (strong) $L_1(X)$-topology, and we have a homogeneous multiplicative process since

$$x(t+h)=x(t)\left\{e+\int_t^{t+h}dy(s)+\underset{t<s_1<s_2<t+h}{\int\!\!\int}dy(s_1)\,dy(s_2)+\ldots\right\}.$$

This is verified by writing down the corresponding relation for the approximating sums and taking limits. Clearly the expression in brackets is independent of $x(t)$ and has the same distribution as $x(h)$. We could write symbolically

$$dx(t)=x(t)\,dy(t)$$

or
$$x(t)=e+\int_0^t x(s)\,dy(s).$$

It is also seen that the x-process is continuous.

Theorem 7.1.1. Let $y(t)$ be a homogeneous stochastic process taking values in the Banach algebra X with the unit element e and satisfying the continuity condition C_1. Then

$$x(t)=e+\int_0^t dy(s)+\underset{0<s_1<s_2<t}{\int\!\!\int}dy(s_1)\,dy(s_2)+\ldots$$

defines a homogeneous multiplicative process.

One may note that if X is a commutative algebra then we get simply

$$x(t)=e+\int_0^t dy(s)+\frac{1}{2}\int_0^t\int_0^t dy(s_1)\,dy(s_2)$$

$$+\frac{1}{6}\int_0^t\int_0^t\int_0^t dy(s_1)\,dy(s_2)\,dy(s_3)+\ldots=\exp\int_0^t dy(s)=\exp y(t).$$

The theorem looks attractive but its usefulness is limited by the continuity condition C_1. Let us now assume that X is a Banach space and a topological algebra with unit element and that $y(t)$ is a homogeneous additive process with values in X and such that $E\|y(t)\|^2<\infty$, so that we can use the $L_2(X)$ topology instead. To define $x(t)$ we start from the same sort of Riemann-Stieltjes sums. In the region $S_r=\{\text{all points }(t_1,t_2,\ldots t_r)\text{ such that }0<t_1<t_2<\ldots<t_r<t\}\subset R^r$ we get rectangles with sides $I_1,I_2,\ldots I_r$ parallell to the

coordinate axes or sums of such rectangles. To such a rectangle ϱ we associate the stochastic element $y(\varrho) = \Delta y(I_1)\Delta y(I_2)\ldots\Delta y(I_r)$ where the order of multiplication may be important. Note that in this way we can define an additive stochastic set function $y(\sigma)$, where σ is a finite union of rectangles, but that this set function does not have independent increments in general. We need a bound for $E\|y(\sigma)\|^2$ and we shall assume that

$$C_2 : E\|y(\sigma)\|^2 \leqslant C^r \cdot \text{volume of } \sigma,$$

where C is some constant and volume means just Lebesgue measure in r-space. Now we can go ahead in the same way as above. It follows that the stochastic integrals in question are uniquely determined. Indeed, the difference between different Riemann-Stieltjes sums can be expressed, as before, in terms of the values of $y(\sigma)$, where the elementary rectangles that form σ together cover only a small part of S_r, and this implies uniqueness as in the previous case. The $L_2(X)$-norm of such an integral is bounded by

$$E\|J_r(t)\|^2 = E\left\|\int_{S_r} dy(t_1)\,dy(t_2)\,\ldots\,dy(t_r)\right\|^2 \leqslant C^r\frac{t^r}{r!}$$

so that
$$\sum_{r=0}^{\infty}\sqrt{E\|J_r(t)\|^2} \leqslant \sum_{r=0}^{\infty}\frac{(Ct)^{r/2}}{\sqrt{r!}} < \infty.$$

This implies convergence in the (strong) $L_2(X)$-topology for the defining series

$$x(t) = \sum_{r=0}^{\infty}J_r(t),\; J_0(t) = e.$$

Let us return for a moment to the new continuity condition C_2. This will have to be verified from case to case, but it may be noted that if the norm in X is introduced via an inner product (x,y) then C_2 can be simplified a bit. Indeed, if $R' = I_1' \times I_2' \times \ldots \times I_r'$ and $R'' = I_1'' \times I_2'' \times \ldots \times I_r''$ are two disjoint rectangles in S_r, then

$$E(y(R'),\, y(R'')) = E(y(I_1')y(I_2')\ldots y(I_r'),\; y(I_1'')y(I_2'')\ldots y(I_r'')).$$

At least for some i we must have $I_i' \cap I_i'' = \phi$. Now if the expected value of the y-process is zero then the above inner product will vanish. In such a case we need only verify C_2 for $\sigma = $ arbitrary rectangle.

Theorem 7.1.2. Let $y(t)$ be a homogeneous multiplicative process taking values in a Banach space X forming a topological algebra with unit element. If $y(t)$ satisfies the continuity condition C_2 then the homogeneous multiplicative process $x(t)$ can be defined uniquely as above.

We are now ready to derive multiplicative limit theorems from the corresponding additive ones. Consider a triangular array of stochastic elements taking values in X

$$\left\{ \begin{array}{l} y_{11}, \\ y_{21}, \; y_{22}, \\ y_{31}, \; y_{32}, \; y_{33}, \\ \quad \cdot \quad \cdot \quad \cdot \quad \cdot \end{array} \right.$$

In each row all the elements are assumed to be independent and to have the same probability distribution. As before we will deal with two cases corresponding to the $L_1(X)$ and $L_2(X)$ methods. We will still need a sort of continuity condition but now expressed in the variables $y_{n\nu}$. In the sums appearing in condition C_1 and C_2 let us replace factors of the type $y(t_\nu) - y(t_{\nu-1})$ by the variables $y_{n\nu}$, and write down the resulting condition. We denote them by C_1' and C_2' respectively. The first one takes the simple form

$$C_1': \sum_{\nu=1}^{n} E\|y_{n\nu}\| \leqslant M < \infty,$$

while the second one is a bit more complicated to write down.

Theorem 7.1.3. Let the $y_{n\nu}$ take values in the Banach algebra X and satisfy C_1'. Let $y(t)$ be an additive homogeneous process satisfying C_1 and such that for every proportion c between 0 and 1 the distributions of the variables

$$\eta_n = \sum_{\nu=1}^{[cn]} y_{n\nu}$$

tend weakly to that of $y(c)$. Then the distributions of the variables

$$\xi_n = (e + y_{n1})(e + y_{n2}) \ldots (e + y_{nn})$$

tend weakly to that of $x(1)$, where $x(t)$ is the homogeneous multiplicative process associated with $y(t)$.

Proof: We have

$$\left\{ \begin{array}{l} \xi_n = e + S_1^{(n)} + S_2^{(n)} + \ldots + S_n^{(n)} \\ x(1) = e + S_1 + S_2 + \ldots S_n + \ldots, \end{array} \right.$$

where

$$\left\{ \begin{array}{l} S_1^{(n)} = \sum_{\nu=1}^{n} y_{n\nu} \\ S_2^{(n)} = \sum_{1 \leqslant \nu < \mu \leqslant n} y_{n\nu} y_{n\mu} \\ \quad \cdot \quad \cdot \quad \cdot \quad \cdot \quad \cdot \quad \cdot \quad \cdot \quad \cdot \end{array} \right.$$

and

$$
\begin{cases}
S_1 = \displaystyle\int_0^1 dy(t) = y(1) \\[2ex]
S_2 = \displaystyle\iint\limits_{0 < s_1 < s_2 < 1} dy(s_1)\,dy(s_2) \\[1ex]
\qquad \cdot \ \cdot \ \cdot \ \cdot \ \cdot \ \cdot \ \cdot \ \cdot \ \cdot \ \cdot
\end{cases}
$$

Working in the L_1 topology over X we know that if we truncate the series for $x(1)$ at the index $n = n_0$ little is changed if n_0 is large enough. We also know that if we replace each term in this finite sum by a Riemann-Stieltjes sum the change is small if the division of the interval $(0,1)$ is fine enough. Now we can do exactly the same thing with the expression for ξ_n; truncate the sum and replace each term in the new sum by sums over blocks corresponding to the division $0 < [nt_1] < [nt_2] < \ldots < n$ of the ν-index. But the distribution of each sum over such a block converges weakly to the corresponding quantity generated by the increments of the y-process. Using the independence of the $y_{n\nu}$ (and of the increments of the y-process) the stated result follows.

We can use the same reasoning to prove the following alternative version.

Theorem 7.1.4. Let the $y_{n\nu}$ take values in the Banach space and topological algebra X and satisfy C_2'. Let $y(t)$ be an additive homogeneous process satisfying C_2. Then the rest of the corresponding statement of the previous theorem holds.

The last two theorems are quite general in character. A more special result is the following.

Theorem 7.1.5. (Multiplicative law of large numbers.) Let y_ν be independent, identically distributed stochastic elements in the Banach algebra, such that $E\|y_1\| < \infty$. Then

$$
\gamma_n = \left(e + \frac{1}{n}\,y_1\right)\left(e + \frac{1}{n}\,y_2\right)\ldots\left(e + \frac{1}{n}\,y_n\right)
$$

converges strongly in probability to the element

$$
\gamma = \exp m, \text{ where } m = Ey_1.
$$

Proof: We have

$$
\gamma_n = e + S_1^{(n)} + S_2^{(n)} + \ldots + S_n^{(n)},
$$

$$\begin{cases} S_1^{(n)} = \dfrac{1}{n} \sum_{1 \leqslant \nu \leqslant n} y_\nu \\[2mm] S_2^{(n)} = \dfrac{1}{n^2} \sum_{1 \leqslant \nu < \mu \leqslant n} y_\nu y_\mu \\[2mm] \text{etc.} \end{cases}$$

where

But we know that with probability one (see section 6.4) $S_1^{(n)} \xrightarrow[\text{strongly}]{} m$. Consider

$$S_2^{(n)} = \frac{1}{n} \sum_{\mu=1}^{n} \frac{\mu-1}{n} S_1^{(\mu)} y_\mu,$$

where

$$S_1^{(\mu)} = \frac{1}{\mu-1} \sum_{\nu=1}^{\mu-1} y_\nu.$$

But with probability one

$$S_1^{(\mu)} = m + \varepsilon_\mu, \; \|\varepsilon_\mu\| \to 0,$$

so that

$$S_2^{(n)} = \frac{m}{n} \sum_{\mu=1}^{n} \frac{\mu-1}{n} y_\mu + \frac{1}{n} \sum_{\mu=1}^{n} \frac{\mu-1}{n} \varepsilon_\mu y_\mu.$$

Hence, with probability one,

$$S_2^{(n)} = \frac{m^2}{2} + \delta_n, \; \|\delta_n\| \to 0.$$

In general, with probability one,

$$S_n^{(k)} \xrightarrow[\text{strongly}]{} \frac{m^k}{k!}.$$

Now we just have to complement this reasoning with a simple uniformity argument. We have

$$\|S_\nu^{(n)}\| \leqslant \frac{1}{n^\nu} \sum_{1 \leqslant k_1 < k_2 < \ldots < k \leqslant n} \|y_{k_1}\| \cdot \|y_{k_2}\| \ldots \|y_{k_\nu}\|$$

so that

$$E\|S_\nu^{(n)}\| \leqslant \frac{(E\|y_1\|)^\nu}{\nu!}.$$

Combining this with the above we have shown that for any $\varepsilon > 0$ we have $P\{\|\gamma_n - \gamma\| > \varepsilon\} \to 0$ as $n \to \infty$, where γ is the element

$$e + m + \frac{m^2}{2} + \frac{m^3}{6} + \ldots = \exp m \in X.$$

(See Notes 7.1.₃.)

7.2. Probabilities on a Banach algebra

With a probability distribution P defined on a Banach algebra X we can study the measure-theoretic properties of derived quantities and concepts in the algebra. The standard algebraic and analytic operations defined in X give rise to stochastic elements of varying degree of complexity. The probability distributions of these elements should be defined, evaluated and studied. When trying to do this we run into many difficulties and here we shall only sketch the rudiments of a theory.

Let x be a stochastic element in X. Introduce the set R of regular elements in the algebra, i.e. those elements that possess an inverse. Then R is an open set (see Notes 7.2.$_1$), which implies that the event that x is regular is a Borel set so that $P(R)$ is defined. If $P(R) > 0$ we can introduce the conditional probability distribution for x^{-1} given that $x \in R$ by

$$P\{x^{-1} \in A \,|\, x \in R\} = \frac{P\{x^{-1} \in A,\, x \in R\}}{P(x \in R)},$$

where A is a Borel set in X. But x^{-1} is a continuous function of x when $x \in R$, so that the event $\{x^{-1} \in A, x \in R\}$ is a Borel set, and we are entitled to speak of the distribution of the inverse x^{-1}.

For a fixed value of λ we can define as usual the resolvent $R(\lambda, \gamma) = (\lambda e - x)^{-1}$ whenever this inverse exists. The set of λ values for which the resolvent does not exist is the spectrum of the element x. In the present context the spectrum is a stochastic set S. It seems likely that the concept of a *stochastic spectrum* will play an important role in the future development of this theory and we should pay some attention to obtaining a satisfactory definition of it. Of course, we can already speak of the probability (for a fixed λ)

$$P(\lambda) = P\{\lambda \in S\} = P\{\lambda e - x \notin R\},$$

and this function $P(\lambda)$ determines some average behavior of the spectrum. Introduce the indicator function $I(\lambda; x)$ of the set $\{x \,|\, \lambda e - x \text{ is singular}\} \subset X$. We have $P(\lambda) = E\{I(\lambda, x)\}$. For any constant c the set $\{(\lambda, x) \,|\, I(\lambda; x) < c\}$ is an open set in the $\Lambda \times X$ space, where Λ stands for the complex λ-plane. This implies that the function $I(\lambda; x)$ is Borel measurable on the product space $\Lambda \times X$ and we can apply Fubini's theorem. In particular the Lebesgue integrals of type

$$\int_{\lambda \in L} I(\lambda; x)\, d\lambda$$

over Borel sets $L \subset \Lambda$ of finite Lebesgue measure exist almost certainly and form stochastic variables. We have proved

Theorem 7.2.1. Let x be a stochastic element in a Banach algebra. The set R of regular elements has a well defined probability $P(R)$. If $P(R) > 0$ we can talk of the (conditional) probability distribution of x^{-1} for x in R. The indicator function $I(\lambda; x)$ of the stochastic spectrum S is Borel measurable on the product space $\Lambda \times X$.

For certain stochastic spectra a *strength function* $s(\lambda)$ can be defined. If the spectra are denumerable point sets on the real line such that the expected number of eigen-values in any given interval (α, β) exists and is $F(\beta) - F(\alpha)$, where $F(x)$ is a differentiable function then the derivative $F'(x)$ is called the strength function because of its quantum mechanical interpretation. We will return to this concept in section 7.5.3.

We can now go ahead and study other quantities associated with a stochastic spectrum. For example the spectral radius

$$r(x) = \sup_{\lambda \in S} |\lambda|$$

or equivalently

$$r(x) = \lim_{n \to \infty} \|x^n\|^{1/n}$$

is a well defined stochastic variable. The easiest way to see this is perhaps to note that $\|x^n\|^{1/n}$ is a stochastic variable for any finite n and these stochastic variables are known to converge for any $x \in X$.

This brings us to a related problem for products of random elements in a Banach algebra. Suppose that $x_1, x_2, ..., x_n$ are independent and identically distributed stochastic elements in X and form the product

$$\gamma_n = x_1 x_2 ... x_n.$$

What can we say about the norm of the product $\|\gamma_n\|$ for large values of n? (See Notes 7.2.$_2$.)

Theorem 7.2.2. Suppose that $x_1, x_2, ..., x_n$ are independent and identically distributed stochastic elements with $E \log^+ \|x_1\| < \infty$. Then the limit

$$r = \lim_{n \to \infty} \frac{1}{n} E \log \|x_1 x_2 ... x_n\|$$

exists and $-\infty \leqslant r < \infty$. If $r \neq -\infty$ then we have almost certainly

$$r = \overline{\lim_{n \to \infty} \frac{1}{n}} \log \|x_1 x_2 ... x_n\|.$$

Proof: Introduce the sequence

$$\alpha_n = E \log \|x_1 x_2 \ldots x_n\|.$$

Either all the α_n are finite or they are all $-\infty$ from some n_0 on. In the first case we can use the subadditive property

$$\alpha_{n+m} = E \log \|x_1 x_2 \ldots x_{n+m}\| \leqslant E \log \{\|x_1 \ldots x_n\| \cdot \|x_{n+1} \ldots x_{n+m}\|\}$$
$$= E \log \|x_1 \ldots x_n\| + E \log \|x_{n+1} \ldots x_{n+m}\| = \alpha_n + \alpha_m$$

and it is known that $1/n\alpha_n$ has a limit (see Notes 7.2.$_3$). This proves the first statement. For the remaining part of the proof we introduce the quantities

$$\xi_n = \frac{1}{n} \log \|x_1 x_2 \ldots x_n\|.$$

If n has the form $r = \nu\mu$ we have

$$\xi_n = \xi_{\nu\mu} = \frac{1}{\nu\mu} \log \|x_1 \ldots x_\nu \, x_{\nu+1} \ldots x_{\nu\mu}\|$$
$$\leqslant \frac{1}{\mu} \left[\frac{1}{\nu} \log \|x_1 \ldots x_\nu\| + \frac{1}{\nu} \log \|x_{\nu+1} \ldots x_{2\nu}\| \right.$$
$$\left. + \ldots + \frac{1}{\nu} \log \|x_{(\mu-1)\nu+1} \ldots x_{\mu\nu}\| \right] \to \frac{1}{\nu} E \log \|x_1 \ldots x_\nu\|$$

almost certainly as $\mu \to \infty$. The right member is close to r if ν is large enough. By a standard argument we can also deal with the values of n that are not of the form $\nu\mu$ and we see that almost certainly $\overline{\lim_n} \, \xi_n \leqslant r$. Now introduce

$$f_n = \frac{1}{n} \sum_1^n \log \|x_i\| - \frac{1}{n} \log \|x_1 \ldots x_n\|.$$

Since $\|x_1 \ldots x_n\| \leqslant \|x_1\| \ldots \|x_n\|$ the variables f_n are non-negative. But

$$E f_n = E \log \|x_1\| - \frac{1}{n} \alpha_n \to E \log \|x_1\| - r$$

and we can now apply Fatou's lemma (see Notes 7.2.$_4$). This states that the variable $f = \overline{\lim_n} \, f_n = E \log \|x_1\| - \overline{\lim_n} \, \xi_n$ has finite expectation and that

$$E f = E \log \|x_1\| - E \overline{\lim_n} \, \xi_n \leqslant \overline{\lim_n} \, E f_n = E \log \|x_1\| - r$$

so that $E \overline{\lim_n} \, \xi_n \geqslant r$. But we already know that with probability one $\overline{\lim_n} \, \xi_n \leqslant r$, so that $\overline{\lim_n} \, \xi_n = r$ almost certainly and this completes the proof.

7.3. Stochastic operators and random equations

Let Z be a Banach space and consider the set X of all bounded linear transformations $X: Z \rightarrow Z$ of Z into Z. With the operator norm the set X forms a Banach algebra. Introducing a probability measure P into X as in the last section it is natural to speak of the elements X as *stochastic operators*. We can study the inverse x^{-1} when it exists, the spectral properties of the operators x and so on.

More generally we can start with a *complete metric space Z*, where we have introduced a metric $\varrho(z_1, z_2)$, and consider a set X of transformations $x: Z \rightarrow Z$. If we introduce a probability measure in X such that for any given $z \in Z$ the element xz is Borel measurable we can still talk of a stochastic operator although we may not be inside the framework of a Banach algebra. Studying resolvents etc. we have to prove various measurability statements but this will not be discussed further in this context; see the Notes for references.

An important special case of the latter type is when X consists of uniformly reducing transformations: there exists a constant $c, 0 < c < 1$, such that for all $x \in X$

$$\varrho(xz_1,\, xz_2) \leqslant c\varrho(z_1, z_2).$$

Consider the *random equation $xz = z$*. Then there is the following stochastic version of the *Banach fixed point theorem*.

Theorem 7.3.1. Let X consist of uniformly reducing transformations of the complete, separable metric space Z into Z. On X we have a probability measure making xz a Borel measurable stochastic element. Then there is a uniquely determined Borel measurable stochastic element ξ with values in Z satisfying the random equation $x\xi = \xi$. This solution can be obtained by iteration starting from an arbitrary Borel measurable stochastic element ξ_1, and putting

$$\left. \begin{aligned} x\,\xi_1 &= \xi_2 \\ x\,\xi_2 &= \xi_3 \\ &\cdot\;\cdot\;\cdot\;\cdot\;\cdot \\ x\,\xi_n &= \xi_{n+1} \\ &\cdot\;\cdot\;\cdot\;\cdot\;\cdot \end{aligned} \right\}$$

We have almost certainly $\xi = \lim_n \xi_n$.

Proof: We have essentially only a question of measurability. Let $z_1, z_2, \ldots$ be an everywhere dense sequence in Z and form, starting from spherical neighborhoods, the subsets

$$A_{in} = \left\{ z \,\middle|\, \varrho(z, z_i) \leqslant \frac{1}{n} \right\} \cap \left\{ \bigcup_{j=1}^{i-1} A_{jn} \right\}^*, \quad i = 1, 2, \dots .$$

Now we define the stochastic operator $x_n z$ when z is a given stochastic element, by

$$x_n z = x z_i \text{ if } z \in A_{in}.$$

But this implies that the event that $x_n z$ takes a value in a Borel set B is equal to the union of the events that $x z_j \in B$ (if $z \in A_{jn}$) for $j = 1, 2, \dots$ Note that the A_{jn} are non-overlapping for a fixed value of n. Hence $x_n z$ and their limit xz have the desirable Borel measurability. Now we can apply this to $x\xi_1, x\xi_2, \dots$, and since their limit, ξ, exists almost certainly because of Banach's fixed point theorem (see Notes $7.3._1$) the statement has been proved.

In this way we arrive at the solution via iterations of the stochastic operator. What happens if we form instead the product of independent stochastic operators similar to the type described above? This question is obviously relevant in many applied problems, but the following theorem should be seen only as a first attempt at solution and sharper versions can be expected.

Theorem 7.3.2. Let Z be a Banach space and let the set X consist of operators mapping Z into Z such that the x's satisfy $\|(xz_1 - xz_2)\| \leqslant \varrho(x)\|z_1 - z_2\|$, where $\varrho(x)$ is a stochastic variable with an expected value ϱ less than one, and such that $E\|x_1 x_2 \dots x_n \xi\|$ is uniformly bounded for any fixed ξ. Draw a sample of independent $x_1, x_2, \dots$ from X and form the stochastic elements $\xi_n = x_1 x_2 x_3 \dots x_n \xi$. As n tends to infinity the stochastic elements ξ_n converge in L_1-mean to a stochastic element ζ, which does not depend upon the initial element ξ.

Proof: Form for arbitrary n and $h > 0$ the difference $\xi_{n+h} - \xi_n = d_{n,h}$. It has the norm

$$\|d_{n,h}\| = \|x_1 x_2 \dots x_n x_{n+1} \dots x_{n+h} \xi - x_1 x_2 \dots x_n \xi\|$$

$$\leqslant \varrho(x_1) \varrho(x_2) \dots \varrho(x_n) \cdot \|x_{n+1} x_{n+2} \dots x_{n+h} \xi - \xi\|$$

as $n \to \infty$, which proves convergence. For two different $\xi_1, \xi_2 \in Z$ we have

$$E\|d_n\| = E\|x_1 x_2 \dots x_n \xi_1 - x_1 x_2 \dots x_n \xi_2\| \leqslant \varrho^n \|\xi_1 - \xi_2\| \to 0$$

so that

$$E\|d_n\| \leqslant \varrho^n \cdot \text{constant} \to 0$$

as $n \to \infty$, so that the limit ζ does not depend upon ξ.

Given a sequence of random operators as above $\dots x_{-1}, x_0, x_1, \dots$ we can form the sequence of stochastic elements

$$\xi_n = \lim_{\nu \to \infty} x_n\, x_{n-1}\, x_{n-2} \dots x_{n-\nu}\, \xi,$$

where ξ is some arbitrary initial element. Then the new sequence ξ_n taking values in Z has a stationary probability distribution and we can say that it is a stationary solution of the *random equation*

$$\xi_{n+1} = x_n \xi_n, n = \dots -1, 0, 1, \dots$$

It may also be noted that it is not essential that Z is a Banach space; this assumption could be weakened and we could work with more general metric spaces.

Very likely the last theorem is capable of generalization and refinements. It seems likely though that some condition is necessary resembling the contraction or reducing properties in order that the variables ξ_n should converge. A related problem that has received some attention is when all the stochastic transformations are measure preserving. This leads us to the so-called *random ergodic theorem* which asserts, not that the ξ_n converge, but that they have a certain average behavior.

Consider a probability space $(\Omega, \mathcal{B}_\Omega, P)$ and let each $\omega \in \Omega$ correspond to a transformation x_ω of another probability space $(Z, \mathcal{B}_z, m)$ onto itself. We shall assume that x_ω is a measure preserving transformation so that $x_\omega(B)$ and $x_\omega^{-1}(B) \in \mathcal{B}_z$ for any $B \in \mathcal{B}_z$ and $m[x_\omega(B)] = m(B)$. The family $\{x_\omega; \omega \in \Omega\}$ should be measurable in the sense that $\{(\omega, z) \mid x_\omega z \in B\} \in \mathcal{B}_\Omega \times \mathcal{B}_z$. Now we introduce in the usual way the new probability space $(\Theta, \mathcal{B}_\Theta, P_\Theta)$ consisting of the sequences $\theta = (\dots \omega_{-1}, \omega_0, \omega_1, \dots)$, where each coordinate ω_n has the probability distribution P and all the coordinates are independent. On Θ we have the shift transformation φ defined so that $\varphi\theta$ means the new θ-sequence whose nth coordinate is θ_{n+1}. Clearly φ is a measure preserving transformation.

Now let $f(z)$ be an m-integrable real valued function defined on Z and form the average

$$S_n = \frac{1}{n} \sum_{\nu=0}^{n-1} f[x_{\omega_{\nu-1}} x_{\omega_{\nu-2}} \dots x_{\omega_0} z].$$

It is appropriate to consider the product measure space $Z \times \Theta$ and to introduce the transformations

$$x(z, \theta) = [x_{\omega_0} z, \varphi\theta].$$

This new transformation is measure preserving and we can apply the (usual) individual ergodic theorem to the space $Z \times \Theta$. Since the νth iterate of $x(z, \theta)$ is given by

$$x^{\nu}(z, \theta) = \nu\text{th iterate of } x(z, \theta) = [x_{\omega_{\nu}-1} \ldots x_{\omega_1} x_{\omega_0} z, \varphi^{\nu}\theta],$$

we have
$$S_n = \frac{1}{n} \sum_{\nu=0}^{n-1} f[x\text{-coordinate of } x^{\nu}(z, \theta)],$$

and we know that almost certainly (on the $Z \times \Theta$-space) the average S_n converges to some integrable limit. Now we apply Fubini's theorem and have proved (see Notes 7.3.$_2$).

Theorem 7.3.3. (*Random ergodic theorem.*) Under the stated conditions there is an event $N \subset \Theta$ of probability zero, $P(N) = 0$, such that if

$$\theta = (\ldots \omega_{-1}, \omega_0, \omega_1, \ldots) \notin N$$

then
$$\lim_{n=\infty} \frac{1}{n} \sum_{\nu=0}^{n-1} f[x_{\omega_{\nu}-1} x_{\omega_{\nu}-2} \ldots x_{\omega_0} z] = \bar{f}(z)$$

exists for almost all $z \in Z$ and $\bar{f}(z) \in L_1(Z)$.

It is not possible in general to claim that the limiting function $\bar{f}(z)$ is almost certainly constant.

7.4. More special structures

If we let a Banach algebra X have more properties we will, of course, be able to say more about its probability distributions. At present this specialization has not yielded anything really interesting, perhaps because of the too detailed and well known structures that one arrives at proceeding by the most direct route.

One such choice would be to require the Banach algebra to possess a fourth operation, division, together with the already introduced operations consisting of addition, multiplication and multiplication by scalars. More precisely, if we demand that X be a (complete) *normed division algebra* so that x^{-1} exists and is a continuous function of x for $x \neq 0$, then X is isomorphic to the complex field (see Notes 7.4.$_1$). But probability distributions on the complex plane can scarcely be expected to give anything really exciting. Take e.g. limit theorems. If we write the complex numbers in polar coordinates, $x = re^{i\varphi}$, then

$$x_1 x_2 \ldots x_n = r_1 r_2 \ldots r_n e^{i(\varphi_1 + \varphi_2 + \ldots + \varphi_n)}.$$

The factors x_r are assumed to have the same distribution and to be independent. Let F denote the simultaneous distribution of $\log r = \varrho$ and φ. We then have to consider addition on the commutative group $R^1 \times T^1$ consisting of the elements (ϱ, φ). The characters are, of course, $e^{iu\varrho + iv\varphi}$, where u is real and v is an integer. Assume that φ does not have a lattice distribution so that $|E \exp iv\varphi| < 1$ for $v \neq 0$. Introduce the Fourier transform

$$\hat{F}(u, v) = \int_{-\infty}^{\infty} \int_{0}^{2\pi} e^{iu\varrho + iv\varphi} \, dF(\varrho, \varphi),$$

and the moments
$$\left.\begin{aligned} m &= E\varrho \\ \sigma^2 &= E(\varrho - m)^2 \end{aligned}\right\}$$

assumed to be finite. Then the variable

$$(P_n, \Phi_n) = \left(\frac{\varrho_1 + \varrho_2 + \dots + \varrho_n - nm}{\sqrt{n}}, \; \varphi_1 + \varphi_2 + \dots + \varphi_n \right)$$

has a probability distribution with the Fourier transform

$$\left[\hat{F}\left(\frac{u}{\sigma\sqrt{n}}, v\right) e^{-\frac{imu}{\sqrt{n}\sigma}} \right]^n.$$

When n tends to infinity

$$\hat{F}\left(\frac{u}{\sigma\sqrt{n}}, v\right) \to \hat{F}(0, v) = E \exp iv\varphi$$

which is less than 1 in absolute value for $v \neq 0$. Hence

$$\left[\hat{F}\left(\frac{u}{\sigma\sqrt{n}}, v\right) \right]^n \to 0 \text{ for } v \neq 0.$$

For $v = 0$ we get, just as in the proof of the simplest version of the central limit theorem

$$\left[\hat{F}\left(\frac{u}{\sigma\sqrt{n}}, 0\right) e^{-\frac{imu}{\sqrt{n}}} \right]^n \to \exp -\frac{1}{2} u^2.$$

This shows that as $n \to \infty$ the asymptotic distribution of (P_n, Φ_n) is normal for P_n, rectangular for Φ_n and the two coordinates are independent. If φ should happen to have a lattice distribution this reasoning can be modified. This case occurs e.g. if X is the real field, $\varphi = 0$, or $\varphi = \pi$.

Another situation close at hand is when the Banach algebra X is *commutative*, especially if it is *semi-simple*. In the latter case we know from

Gelfand's theory that X is isomorphic to the family of continuous functions $c(t)$ on a certain compact set T (see Notes 7.4.$_2$). Again the situation presents little of interest that we have not already met before. To study limit theorems e.g. we will consider products $c_1(t)c_2(t)...c_n(t)$ and sums $c_1(t)+c_2(t)+...+c_n(t)$ leading respectively to the logarithmiconormal and uniform (perhaps on some subgroup) distributions or to the normal distribution. This is when we consider the value of the product or sum for each t separately, but we can also determine the simultaneous asymptotic distribution for any finite number of t's or on the σ-algebra generated by the corresponding cylindersets.

7.5. Illustrations

7.5.1. A very important stochastic algebra consists of *random $k \times k$ matrices*. It is sufficiently rich in structure to support some useful and substantial probability theory. We shall now discuss some aspects of it. If we write $x \in X$ as

$$x = \{x_{ij};\ i,j = 1,\ 2,...k\}$$

and introduce the norm

$$\|x\| = \max_i \sum_j |x_{ij}|,$$

then X is a Banach algebra.

Supposing the stochastic variables x_{ij} to have existing mean values m_{ij} then, of course, x also has a mean value m

$$m = \{m_{ij};\ i,j = 1,2,...k\}$$

If $x_1, x_2,...$ are independent, stochastic matrices in X with mean values $m_1, m_2,... \in X$, then obviously the product $x_1 x_2...x_n$ has the mean value $m_1 m_2...m_n$. This statement holds for an arbitrary Banach algebra. Indeed, if y and η are two stochastically independent elements in some Banach algebra X with mean values m and μ, then, putting $z = y\eta$, the stochastic variable $\|z\| \leqslant \|y\| \cdot \|\eta\|$ is integrable (z is, of course, Borel measurable). $Ez = \nu$ then exists and we should have for any bounded linear functional $x^* \in X^*$

$$x^*(\nu) = Ex^*(z) = \int\int x^*(y\eta)\,P_1(dy)\,P_2(d\eta).$$

But $x^*(z) = x^*(y\eta) = x_y^*(\eta)$ is a bounded linear functional x_y^* of η for fixed y. Fubini's theorem then gives us

$$x^*(\nu) = \int x_y^*(\mu) P_1(dy) = \int x^*(y\,\mu) P_1(dy) = x^*(m\,\mu)$$

so that $\nu = Ey\eta = m\mu = Ey \cdot E\eta$ as stated.

We have in particular $Ex_1 x_2 \ldots x_n = m^n$ if all the factors x_ν have the same mean value m. In the matrix case it is well known that the behavior of m^n for large values of n is related to the properties of the largest eigenvalues of m (use e.g. Jordan's canonical form). In the general case it depends in a less simple way upon the spectral properties of the element m.

Now consider $y_n = \{y_{ij}^{(n)}; \; i, j = 1, 2, \ldots k\}$ with $y_n = x_1 x_2 \ldots x_n$. Then, assuming second order moments to exist,

$$Ey_{ij}^{(n)} y_{\alpha\beta}^{(n)} = \sum Ex_{ii_1}^{(1)} x_{\alpha\alpha_1}^{(1)} \, Ex_{i_1 i_2}^{(2)} x_{\alpha_1 \alpha_2}^{(2)} \ldots Ex_{i_{n-1}j}^{(n)} x_{\alpha_{n-1}\beta}^{(n)},$$

with summation extended over $\alpha_1, \alpha_2, \ldots \alpha_{n-1}, i_1, i_2 \ldots i_{n-1}$, or if we introduce the $k^2 \times k^2$ covariance matrices

$$\begin{cases} C = \{C_{ij, \alpha\beta}; \; i, j, \alpha, \beta = 1, 2, \ldots k\} \\ C_{ij, \alpha\beta} = Ex_{ij}^{(1)} x_{\alpha\beta}^{(1)}, \end{cases}$$

we get
$$\{Ey_{ij}^{(n)} y_{\alpha\beta}^{(n)}\} = C^n.$$

Again it is possible to make asymptotic statements. We can also continue with higher order moments.

While this sort of computation is not without interest it does not help much as far as limit theorems are concerned. This is obvious already on the real line. Indeed if $X = R^1$ then we know the asymptotic behavior of $y_n = x_1 x_2 \ldots x_n$ which can be expressed in terms of the logarithmico-normal distribution. If $a = E \log x$ and $b = \mathrm{Var}\,(\log x)$ exist then $(\log y_n - na)(nb)^{-1/2}$ is asymptotically normal. We would then expect y_n to be of the order e^{na} while $Ey_n = (Ex)^n$; the first corresponds to a geometric mean and the second to an arithmetic one and they will differ substantially in general.

We know from section 7.1 that in the study of certain limit problems we should consider related multiplicative homogeneous processes depending upon a continuous time parameter t. Let us consider two such processes in the matrix case.

Let $y(t)$ be an additive homogeneous stochastic process, $t \geqslant 0$, taking as values $k \times k$-matrices and behaving like a Poisson process so that

$$y(t+h) = \begin{cases} y(t) + hA \text{ with probability } 1 - \lambda h + o(h) \\ y(t) + hA + B \text{ with probability } \lambda h + o(h), \end{cases}$$

or more precisely we start from an ordinary Poisson point process with the timepoints t_ν of events generated with a Poisson density λ. Between two such t_ν the process changes linearly proportionately to the constant matrix A. At a t_ν the y-process jumps by the also constant matrix B. This determines the probability distribution of the process. It satisfies condition C_1 of 7.1 since

$$\sum E\|y(t_\nu) - y(t_{\nu-1})\| \leqslant \sum (t_\nu - t_{\nu-1})\,[\|A\| + \lambda\|B\| + o(1)] = O(1).$$

Hence the corresponding multiplicative homogeneous process $x(t)$ is determined. If A and B commute $x(t)$ is simply $x(t) = \exp[tA + n(t)B]$, where $n(t)$ is a Poisson distributed stochastic variable, but if $AB \neq BA$ the x-process is more complicated. Anyway we can compute its mean values and covariances easily enough. Putting

$$M(t) = \{m_{ij}(t); \ i,\ j = 1,2,\ldots k\},\, m_{ij}(t) = Ex_{ij}(t),$$

we get

$$M(t+h) = Ex(t+h) = M(t)[I + hA + h\lambda B + o(h)]$$

so that

$$M'(t) = M(t)(A + \lambda B)$$

and

$$M(t) = \exp t(A + \lambda B).$$

To determine the covariances introduce the $k^2 \times k^2$-matrix

$$C(t) = \{c_{ij.\,\alpha\beta}(t); \ i, j, \alpha, \beta = 1, 2 \ldots k\} = \{\operatorname{cov}[x_{ij}(t), x_{\alpha\beta}(t)]\},$$

where we have ordered the indices in some order, say lexicographic. If we denote the Kronecker product of two matrices R and S by $R \times S$ we have

$$C(t) = \{Ex_{ij}(t)\,x_{\alpha\beta}(t) - m_{ij}(t)\,m_{\alpha\beta}(t)\} = Ex(t) \times x(t) - M(t) \times M(t).$$

But the matrices $S(t) = Ex(t) \times x(t)$ satisfy the relation

$$S(t+h) = Ex(t+h) \times x(t+h) = E[x(t) + x(t)\Delta y] \times [x(t) + x(t)\Delta y] + \text{terms of}$$

smaller order, where we have put $\Delta y = y(t+h) - y(t)$. Hence

$$S(t+h) = S(t) + hEx(t) \times x(t)[A + \lambda B]$$
$$+ hEx(t)[A + \lambda B] \times x(t) + Ex(t)\Delta y \times x(t)\Delta y + \ldots$$

so that

$$S'(t) = S(t)(I \times [A + \lambda B]) + S(t)([A + \lambda B] \times I) + \lambda S(t)\,B \times B$$

and

$$S(t) = \exp t\{I \times (A + \lambda B) + (A + \lambda B) \times I + \lambda B \times B\}.$$

It is possible to find the asymptotic behavior of $M(t)$ and $S(t)$ for large values of t as well as to extend this to moments of higher order.

Now we will start from another y-process. Assume that $y(t)$ has normal distributions with independent increments and with mean values zero. The covariances will be proportional to t and we write

$$Ey_{ij}(t)\,y_{\alpha\beta}(t) = t\,b_{ij.\alpha\beta}.$$

An appropriate definition of the norm is here given by

$$\left.\begin{array}{l}\|U\|^2 = (U,\,U) \\ (U,\,V) = \sum_{i,j} u_{ij}\,v_{ij},\end{array}\right\}$$

where U and V are real $k \times k$ matrices with the elements u_{ij} and v_{ij} respectively. To verify condition C_2 we need only consider $y(R)$, where R is a rectangle in S_r, see section 7.1. But

$$\|y(R)\|^2 = \sum y^{(1)}_{i_1 i_2} y^{(2)}_{i_2 i_3} \cdots y^{(r)}_{i_r i_{r+1}} y^{(1)}_{j_1 j_2} y^{(2)}_{j_1 j_2} \cdots y^{(r)}_{j_r i_{r+1}},$$

where the sum is extended over all values between 1 and k of the indices and where $y^{(\nu)}_{ij}$ is the $(i,\,j)$-element of the matrix consisting of the increment of $y(t)$ over the νth side of length Δ_ν of the rectangle R. This implies

$$E\|y(R)\|^2 = \sum Ey^{(1)}_{i_1 i_2} y^{(1)}_{i_1 j_1} \cdot Ey^{(2)}_{i_2 i_3} y^{(2)}_{j_1 j_2} \cdots Ey^{(r)}_{i_r i_{r+1}} y^{(r)}_{j_r i_{r+1}}$$

$$= O(k^{2r+1}\,b^r\,\Delta_1\,\Delta_2 \cdots \Delta_r) = 0\,(\text{const}^r \cdot m(R)),$$

where $b = \max|b_{ij\alpha\beta}|$ and this shows that our condition is satisfied so that the corresponding multiplicative homogeneous process $x(t)$ is well defined.

The infinitesimal covariances of $x(t)$ are $\lim\limits_{h \to 0} 1/h$ {conditional covariance $[x_{ij}(t+h) - x_{ij}(t),\, x_{\alpha\beta}(t+h) - x_{\alpha\beta}(t)]\} = \sum_{l,\lambda=1}^{k} x_{il}(t)\,x_{\alpha\lambda}(t)\,b_{lj.\lambda\beta} = q_{ij.\alpha\beta}\,[x(t)]$, where the conditional covariance is taken for a fixed value of $x(t)$. The $q_{ij.\alpha\beta}(x)$ are quadratic forms in the elements of the x-matrix and the $k^2 \times k^2$ matrix $q = \{q_{ij.\alpha\beta}\}$ can be written as $[x(t) \times x(t)]B$. The Fokker-Planck equation is then

$$\frac{\partial p}{\partial t} = \frac{1}{2} \sum \frac{\partial^2}{\partial x_{ij}\,\partial x_{\alpha\beta}} [q_{ij.\alpha\beta}(x)\,p]$$

and appropriate initial values should be prescribed for its solution. In the present case these would correspond to $P\{x(0) = I\} = 1$.

If the mean values of the y-process do not vanish but are

$$Ey(t) = tA, A = \{a_{ij}\},$$

then the above should be modified to

$$\frac{\partial p}{\partial t} = \frac{1}{2} \sum \frac{\partial^2}{\partial x_{ij}\,\partial x_{\alpha\beta}} [q_{ij,\alpha\beta}(x)p] - \sum \frac{\partial}{\partial x_{ij}} [l_{ij}(x)p],$$

where $L = \{l_{ij}(x)\}$ and $L = x(t)A$ so that the $l_{ij}(x)$ are linear forms in the x_{ij}.

To find the expected values of the x-process we perform a partial integration and get

$$\frac{\partial}{\partial t} Ex_{ij}(t) = \frac{\partial}{\partial t} m_{ij}(t) = El_{ij}(x)$$

so that
$$\frac{\partial m(t)}{\partial t} = Ex(t)A = m(t)A; \; m(t) = \exp tA.$$

Similarly for the second order moments we get

$$\frac{\partial}{\partial t} c_{ij,\alpha\beta}(t) = E q_{ij,\alpha\beta}[x(t)] + El_{ij}(x)x_{\alpha\beta}(t) + El_{\alpha\beta}(x) x_{ij}(t)$$

or
$$\frac{\partial c(t)}{\partial t} = c(t)B + c(t)[A \times I + I \times A]$$

so that
$$c(t) = \exp t\{B + A \times I + I \times A\}.$$

Compare with the similar expressions earlier in this section.

The x-process studied above is of interest in the study of many limit problems. It is therefore a task of considerable importance to find methods enabling us to handle the solutions of these Fokker-Planck equations. Of course, it may be possible to solve the equations numerically by a straightforward approach, but this would be of limited practical value except in special cases, since the great number of parameters make tabulation difficult. Indeed we have k^2 parameters for the infinitesimal mean values and $\frac{1}{2}k^2(k^2+1)$ parameters for the infinitesimal covariances, in all $N = \frac{1}{2}k^2(k^2+3)$ parameters. For $k=2$ this is $N=14$ and already for $k=3$ we get the forbidding number of $N=54$ parameters. This makes it almost necessary to use an analytic or approximate approach, say of representing the solution in terms of tabulated functions or by using some suitable series expansion or integral representation. This problem obviously requires further study.

Before leaving this problem let us make some qualitative remarks. For simplicity let us consider stochastic 2×2 matrices M_h with vanishing mean values operating upon the vectors x in R^2, $\Delta x = x(t+h) - x(t) \cong M_h x(t)$. The distribution of $\Delta x \in R^2$ is determined by the infinitesimal covariances

$$q_{ij}(x) = \operatorname{cov}\left[(Mx(t))_i, (Mx(t))_j\right]$$

evaluated for a certain $x(t) = x$, and where M is a stochastic 2×2 matrix. If the matrix $Q(x) = \{q_{ij}(x); \, i,j = 1,2\}$ is non-singular at x, then diffusion takes place in all directions from the point x. If we require that the probability should be restricted to some subset of R^2, the matrix $Q(x)$ must be singular for certain values of x, say in a subset S. But M has four elements so that we can represent it as

$$M = \sum_{\nu=1}^{r} \xi_\nu A_\nu,$$

where the A_ν are non-stochastic 2×2 matrices, the ξ_ν are uncorrelated stochastic variables with variance one and $r \leqslant 4$. Then

$$q_{ij}(x) = \sum_{\nu=1}^{r} (A_\nu x)_i (A_\nu x)_j$$

so that if z is an arbitrary column vector

$$z^* Q(x) z = \sum_{\nu=1}^{r} (A_\nu x, z)^2.$$

Now if this quadratic form vanishes for a certain non-trivial vector z, then no diffusion takes place in the direction of the vector z. Let S be the set of x values in R^2 where $Q(x)$ becomes singular

$$S = \{x \,|\, \det Q(x) = 0\},$$

so that in S we have

$$0 = \det Q(x) = \sum_{\nu,\mu} \begin{vmatrix} (A_\nu x)_1 (A_\mu x)_1 & (A_\nu x)_1 (A_\mu x)_2 \\ (A_\nu x)_2 (A_\mu x)_1 & (A_\nu x)_2 (A_\mu x)_2 \end{vmatrix} = \sum_{\nu<\mu} \begin{vmatrix} (A_\nu x)_1 & (A_\mu x)_1 \\ (A_\nu x)_2 & (A_\mu x)_2 \end{vmatrix}^2$$

which implies that

$$\begin{vmatrix} (A_\nu x)_1 & (A_\mu x)_1 \\ (A_\nu x)_2 & (A_\mu x)_2 \end{vmatrix} = 0$$

for all ν and μ. But this is a set of homogeneous equations of the second order in x so that S can be the isolated point $x = 0$, one or two straight lines or the entire plane.

First consider $r = 1$. Then $S = R^2$ and the quadratic form $z^* Q(x) z$ vanishes if $(Ax, z) = 0$. Diffusion takes place only in a direction parallel to the vector Ax so that we are led to the differential equation for the *singular curves*

$$\frac{dx_2}{dx_1} = \frac{(Ax)_2}{(Ax)_1}$$

which is treated in all elementary text books on differential equations. Its solutions form families of curves C_α consisting of straight lines, ellipses, spirals, hyperbolas and parabolas of various types. If the initial distribution is concentrated on a C_α then the mass never leaves this curve but can only diffuse along it. If A is singular we get a still more singular situation since if the vector x annihilates $A_1, A_1 x = 0$, then the mass starting out from x stays there indefinitely.

Now consider $r=2$. Then S consists of the points, where

$$\det A\,(x) = \begin{vmatrix} (A_1 x)_1 & (A_2 x)_1 \\ (A_1 x)_2 & (A_2 x)_2 \end{vmatrix} = 0$$

implying $(A_1 - \lambda A_2)x = 0$ for some scalar λ. For a non-trivial vector x we must have $\det\,(A_1 - \lambda A_2) = 0$ so that only two different values of λ are possible. Excluding the case that A_1 and A_2 are multiples of each other $(A_1 = \lambda A_2)$, which corresponds to the case $r=1$, we see that $(A_1 - \lambda A_2)x = 0$ has solutions $x = \alpha u$ where $u \in R^2$ and α is an arbitrary scalar. When is such a straight line singular in the way described? If the only possible diffusion is in the direction u, then we must have

$$0 = (A_1 x_1 n)^2 + (A_2 x_1 n)^2 = \alpha^2 (A_1 u_1 n)^2 + \alpha^2 (A_2 u_1 n)^2,$$

where n is a vector orthogonal to u, so that

$$\left.\begin{aligned} A_1 u = \mu_1 u \\ A_2 u = \mu_2 u \end{aligned}\right\}.$$

This means that A_1 and A_2 should have one or two right eigen-vectors in common and we get one or two singular straight lines.

For $r=3$ or 4 we get similar answers. Singular curves occur only if the A_ν matrices have eigenvectors in common; otherwise the probability mass fills out the entire plane ($x = 0$ is of course always a singular point).

Let $E \subset R^2$ be a region the finite part of which is bounded by singular curves. If the initial distribution is contained in E then it is intuitively clear that no mass will ever leave E, $P_t(E) \equiv 1$. To show this let us for simplicity assume that E is bounded by two rays L_1 and L_2 extending from the origin. It is convenient to use polar coordinates ϱ and θ and we let L_1 and L_2 correspond to the angles θ_1 and θ_2. It is easy to write down the Fokker-Planck equation

$$\frac{\partial}{\partial t} p(\theta, \varrho; t) = \frac{1}{2} \frac{\partial^2}{\partial \theta^2} [k_1(\theta) p] + \frac{\partial^2}{\partial \theta \partial \varrho} [\varrho \, k_2(\theta) p] + \frac{1}{2} \frac{\partial^2}{\partial \varrho^2} [\varrho^2 k_3(\theta) p],$$

where $k_i(\theta)$ are second order trigonometric polynomials in θ and $k_1(\theta)$ and $k_2(\theta)$ vanish for $\theta = \theta_1$ and θ_2. Now we can integrate over E

$$P_t(E) = \int_{\theta_1}^{\theta_2} \int_0^\infty \varrho p(\theta, \varrho; t) d\theta d\varrho$$

and after a few partial integrations we get $(\partial/\partial t) P_t(E) = 0$ as stated. We have used the fact that $k_1(\theta)$ has second order zeroes and $k_2(\theta)$ first order zeroes at $\theta = \theta_1$ and θ_2.

We now turn to a different aspect of the same problem: the behavior of the angle $\theta(t)$ for large values of t. Assume there are no singular lines, $k_1(\theta) > 0$. We shall no longer ignore the infinitesimal mean value of θ. Neither shall we identify angles differing by multiples of 2π. Instead we follow $\theta(t)$ as it varies continuously in t; see also 4.4.1. Note that $\theta(t)$ forms a Markov process just as $(\theta(t), \varrho(t))$ does, since $\Delta x = M_h x$ is a homogeneous function of x. The Fokker-Planck equation for the frequency function $\pi(\theta; t)$ of *this* angle $\theta(t)$ can be written

$$\frac{\partial \pi(\theta; t)}{\partial t} = \frac{1}{2} \frac{\partial^2}{\partial \theta^2} [k_1(\theta) \pi] - \frac{\partial}{\partial \theta} [k_4(\theta) \pi],$$

where $k_4(\theta)$ is a first order trigonometric polynomial. This could also have been derived from the earlier Fokker-Planck equation (with mean value terms added) by integrating out the variable ϱ. We have

$$\frac{d}{dt} E \theta(t) = \int_{-\infty}^\infty \theta \frac{\partial \pi(\theta: t)}{\partial t} d\theta = \int_{-\infty}^\infty \pi(\theta; t) k_4(\theta) d\theta = \int_0^{2\pi} p(\theta; t) k_4(\theta) d\theta,$$

where the last relation follows since $k_4(\theta)$ is periodic with the period 2π. Suppose we start off the process with the (existing) stationary frequency function $p(\theta)$. To compute $p(\theta)$ we have to solve

$$\frac{1}{2} \frac{\partial^2}{\partial \theta^2} [k_1(\theta) p] - \frac{\partial}{\partial \theta} [k_4(\theta) p] = 0$$

or

$$\frac{1}{2} \frac{\partial}{\partial \theta} [k_1(\theta) p] - k_4(\theta) p = a$$

which has the solutions

$$p(\theta) = \frac{1}{k_1(\theta)} \left[b + 2a \int_0^\theta \exp\left(-2 \int_0^u \frac{k_4(v)}{k_1(v)} dv \right) du \right] \exp 2 \int_0^\theta \frac{k_4(u)}{k_1(u)} du,$$

where the constants a and b should be chosen so that $p(\theta)$ is a frequency function satisfying the boundary condition $p(2\pi) = p(0)$. But

$$
\begin{cases}
p(0) = \dfrac{b}{k_1(0)} \\[2ex]
p(2\pi) = \dfrac{b + 2aI_1}{k_1(0)} \exp I_2,
\end{cases}
$$

where

$$
\begin{cases}
I_1 = \displaystyle\int_0^{2\pi} \exp\left(-2\int_0^u \frac{k_4(v)}{k_1(v)}\,dv\right) du > 0 \\[3ex]
I_2 = 2\displaystyle\int_0^{2\pi} \frac{k_4(u)}{k_1(u)}\,du.
\end{cases}
$$

This gives us the conditions

$$
b = (b + 2aI_1)\exp I_2
$$

so that

$$
a = \frac{1 - \exp I_2}{2\,I_1 \exp I_2}\,b.
$$

Choosing b as some positive number we can make $p(\theta)$ positive, since $b + 2aI_1 = b\exp(-I_2) > 0$, and with integral 1 over $(0, 2\pi)$. Combining the above we see that

$$
\frac{d}{dt}\,E\theta(t) = \int_0^{2\pi} p(\theta)\,k_4(\theta)\,d\theta = -2\pi a
$$

so that

$$
E\,\frac{\theta(t) - \theta(0)}{t} = -2\pi a.
$$

Or in words: the point circles the origin with an average angular velocity of a revolutions (in the negative direction) per time unit.

This example may be of some general interest when looked at from the point of view of section 4.5.1.

7.5.2. Now let us assume that all the possible elements of the matrices x 1) are positive and 2) satisfy an inequality of the form

$$
\frac{\max x_{ij}}{\min x_{ij}} \leqslant C < \infty.
$$

For the product $y^n = \{y_{ij}^n\} = x^1 x^2 \ldots x^n$ we have

$$
\min_{i,j} y_{ij}^n \leqslant \|y^n\| \leqslant k \max_{i,j} y_{ij}^n,
$$

where we use the same norm as in the beginning of section 7.5.1. But

$$\frac{y_{ij}^n}{y_{\alpha\beta}^n} = \frac{\sum\limits_{\nu,\mu} x_{i\nu}^n z_{\nu\mu} x_{\mu j}^1}{\sum\limits_{\nu,\mu} x_{\alpha\nu}^n z_{\nu\mu} x_{\mu\alpha}^1} \leqslant C^2,$$

where $\{z_{\nu\mu}\} = x^2 x^3 \ldots x^{n-1}$, so that

$$\max_{i,j} y_{ij}^n \leqslant C^2 \min_{i,j} y_{ij}^n.$$

Hence

$$\frac{1}{n} \log \min_{i,j} y_{ij}^n \leqslant \frac{1}{n} \log \|y^n\| \leqslant \frac{1}{n} \log kC^2 + \frac{1}{n} \log \min_{i,j} y_{ij}^n.$$

But the middle term tends almost certainly to a limit r (if r is finite; see Notes 7.1 and section 7.1). This and above inequality implies that

$$\frac{1}{n} \log y_{ij}^n \to r$$

almost certainly (if r is finite) for all indices i and j.

It has also been shown (see Notes 7.5.2) that if $E|\log x_{11}|^{2+\delta} < \infty$, for some positive δ, then there exist real constants a and b such that

$$\lim_{n \to \infty} P\left\{\frac{\log y_{ij}^n - na}{\sqrt{nb}} \leqslant x\right\} = \frac{1}{\sqrt{2\pi}} \int_{-\infty}^{x} e^{-\frac{z^2}{2}} dz$$

if b is not zero. If b is zero then

$$\frac{\log y_{ij}^n - na}{\sqrt{n}} \to 0$$

in probability as $n \to \infty$.

It is not clear how typical this fact (asymptotic normality) is, but it is obvious that it does not hold for matrices with general entries. This is seen already for orthogonal matrices where we know the relevant limit theorems, see Chapter 3. Anyway it would be desirable to put this result in correspondence with the general theory.

7.5.3. To illustrate the concept of a random spectrum let us mention two special cases.

The first is from *multivariate statistical* analysis. In this theory much attention is given to certain random matrices of estimated variances, covariances and correlation coefficients. The entries will have Wishart or sometimes normal distributions. The eigenvalues $\lambda_1, \lambda_2, \ldots \lambda_p$ are used in the statistical analysis and they form stochastic variables with a simultaneous probability distribution whose frequency function can sometimes be evaluated explicitly. Note that we must be able to identify the various eigenvalues in some way, e.g. by ordering them $\lambda_1 \geqslant \lambda_2 \geqslant ... \geqslant \lambda_p$. As an example of such a frequency function assume that the x_{ij} are independent except for a symmetry condition so that x_{ij} and $x_{\alpha\beta}$ are independent if $(i,j) \neq (\alpha,\beta)$ and $(i,j) \neq (\beta,\alpha)$, $x_{ij} = x_{ji}$, and that $x_{ij} = N(0, \sigma_{ij}^2)$ with $\sigma_{ij}^2 = \frac{1}{2}$ if $i \neq j$, $\sigma_{ii}^2 = 1$. Then the joint frequency function of the eigenvalues is given by the expression

$$\frac{\exp - \dfrac{1}{2} \sum_1^p \lambda_\nu^2}{2^{p/2} \prod_1^p \Gamma \left(\dfrac{p+1-i}{2} \right)} \prod_{j<j} (\lambda_i - \lambda_j)$$

in the region $\lambda_1 \geqslant \lambda_2 \geqslant ... \geqslant \lambda_p$ and 0 otherwise. It is possible to evaluate the distribution of simple functions of the eigenvalues, such as the determinant $= \lambda_1 \lambda_2 ... \lambda_p$, see Notes 7.5.3.

The form of the above frequency function is not easy to handle if p is large. It is therefore interesting to note the following asymptotic result (see Notes 7.5.3). Let the x_{ij} be independent except for symmetry but with distributions about which we assume only that 1) they are symmetric around zero, 2) they have variance 1 and 3) they have uniformly bounded moments of any finite order $|Ex_{ij}^r| \leqslant A_r < \infty, r = 1, 2, ...$. Then consider the symmetric random matrices

$$M_n = \frac{1}{\sqrt{n}} \{x_{ij}; \, i, j = 1, 2, ..., n\}$$

with the real eigen-values $\lambda_1, \lambda_2, ... \lambda_n$. Let us form the stochastic variables

$$\mu_p^{(n)} = \frac{1}{n} \sum_{\nu=1}^n \lambda_\nu^p = \frac{1}{n} tr M_n^p = \frac{1}{n^{1+p/2}} \sum x_{i_1 i_2} x_{i_2 i_3} ... x_{i_p i_1},$$

where all the indices $i_1, i_2, ..., i_p$ run through the numbers $1, 2, ..., n$. It is easily seen that $E\mu_p = 0$ if p is odd because of the symmetry of the distributions. If p is even, say $p = 2\nu$, then

$$E \mu_{2\nu}^{(n)} = \frac{1}{n^{1+\nu}} \sum Ex_{i_1 i_2} x_{i_2 i_3} ... x_{i_{2\nu} i_1}.$$

It is clear that many terms in the above sum will be zero since each factor has mean value zero and because of the independence assumption. It is not difficult to prove that for large values of n the leading term in the asymptotic expansion of the sum will be of the form $c_{2\nu}\cdot n^{\nu+1}$ and the remainder of smaller order. Hence $\lim_{n\to\infty} E\mu_{2\nu}^{(n)} = c_{2\nu}, \nu = 1, 2, \ldots$ To compute the c's is a more tricky combinatorial problem. Using a recursive argument it has been shown (see Notes 7.5.3) that $c_{2\nu} = (2\nu)!/\nu!\,(\nu+1)!$. But the characteristic function with these moments (and the odd ones vanishing) is the entire function

$$\sum_{\nu=0}^{\infty} c_{2\nu}\frac{(iz)^{2\nu}}{(2\nu)!} = \sum_{\nu=0}^{\infty}\frac{(-z^2)^{\nu}}{\nu!\,(\nu+1)!} = \frac{1}{2}J_1(z) = \frac{2}{\pi}\int_{-1}^{1} e^{2izu}\sqrt{1-u^2}\,du.$$

But this implies that for any polynomial $c(x)$ we have

$$\lim_{n\to\infty} E\frac{1}{n}\sum_{\nu=1}^{n} c(\lambda_\nu) = \frac{1}{\pi}\int_{-2}^{2} c(u)\sqrt{1-\frac{u^2}{4}}\,du.$$

Now one could try to use this to show that for any interval $(\alpha, \beta)\subset(-2, 2)$ we have

$$\lim_{n\to\infty} E\left[\frac{1}{n}\text{ number of }\lambda_\nu\in(\alpha,\beta)\right] = \frac{1}{\pi}\int_{\alpha}^{\beta}\sqrt{1-\frac{u^2}{4}}\,du.$$

The strength function (see 7.2) would then have limit

$$s(u) = \frac{1}{\pi}\sqrt{1-\frac{u^2}{4}}$$

and would not depend upon the form of the distribution of the x's as long as the stated assumptions are satisfied. See Notes 7.5.3.

We will instead use the following argument that describes not only the average behavior of the spectrum but tells us something about the actual asymptotic form of the stochastic spectrum. To prove this more informative statement let us consider not only the average behavior but also the variances of the quantities $\mu_{2\nu}^{(n)}$. We have

$$E(\mu_{2\nu}^{(n)})^2 = \frac{1}{n^{2+2\nu}}\sum Ex_{i_1 i_2}x_{i_2 i_3}\ldots x_{i_{2\nu} i_1}x_{j_1 j_2}x_{j_2 j_3}\ldots x_{j_{2\nu} j_1}.$$

To evaluate this sum let us fix the indices $i_1, i_2, \ldots i_{2\nu}$ and see what happens when the other ones vary. But then the terms factor into

$$E x_{i_1 i_2}x_{i_2 i_3}\ldots x_{i_{2\nu} i_1}\cdot E x_{j_1 j_2}x_{j_2 j_3}\ldots x_{j_{2\nu} j_1}$$

except if at least one of the pairs $(j_1, j_2), (j_2 j_3), \ldots, (j_{2\nu}, j_1)$ is equal to at least one of the pairs $(i_1, i_2), (i_2 i_3), \ldots, (i_{2\nu}, i_1), (i_2, i_1), (i_3, i_2), \ldots, (i_{2\nu}, i_1)$. But the latter situation occurs only in a negligible number of cases so that actually

$$\lim_{n\to\infty} E\,(\mu_{2\nu}^{(n)})^2 = c_{2\nu}^2$$

so that $\mathrm{Var}\,(\mu_{2\nu}^{(n)}) \to 0$ and all the moments $\mu_{2\nu}^{(n)}$ converge in the mean to the corresponding limits $c_{2\nu}$. This also holds for the moments of odd order. Now let us consider the distribution function $F^{(n_i)}(u)$ for an arbitrarily fixed value u and for an arbitrary subsequence $n_1, n_2, n_3, \ldots$ of the natural numbers. By $F^{(n)}(u)$ we mean the distribution function with the jump $1/n$ in each of the points $\lambda_1, \lambda_2, \ldots \lambda_n$. But all the $\mu_{2\nu}^{(n)}$ converge in probability, so that from $\{n_i\}$ we can select a subsequence $\{n_i'\}$ so that the $\mu_2^{(n_i')}$ converge almost certainly, then a subsequence $\{n_i''\} \subset \{n_i'\}$ so that the $\mu_4^{(n_2'')}$ converge almost certainly etc. The diagonal procedure gives us a sequence $n_1', n_2', \ldots$ for which all the moments converge almost certainly. But the limiting moments correspond, as we have seen, to an entire characteristic function, which implies that $F^{(k)}(u)$ converges almost certainly to $F(u)$ when k runs through a certain subsequence of $\{n_i\}$. Since we can do this starting with any given sequence $n_1, n_2, \ldots$ it follows that $F^{(n)}(u)$, $n = 1, 2, \ldots$ converges in probability to $F(u)$. The stochastic spectrum tends in the present case to a fixed spectrum. This may seem surprising but seems to be an ergodic property connected with the independence of the distributions of the entries of the matrix. If this is so we could expect this sort of convergence to hold much more generally.

7.5.4. Consider the rational linear transformation $1/(\alpha - x)$, $-1 < x < 1$, from the point of view of section 7.3. The parameter α will be given a probability distribution in the set $|\alpha| > 1$. Since

$$\left| \frac{1}{\alpha - x_1} - \frac{1}{\alpha - x_2} \right| \leqslant \frac{|x_1 - x_2|}{(|\alpha| - 1)^2}$$

we can choose the quantity $\varrho(\alpha) = 1/(|\alpha| - 1)^2$ and require that $E\varrho(\alpha) < 1$. This is certainly true if $|\alpha| > 2$ with probability one.

The present situation does not correspond exactly to Theorem 7.3.2 (z is not a Banach space here), but it can be shown by a similar reasoning as in 7.3 that we can form the convergent continued fraction

$$y_n = \cfrac{1}{\alpha_n - \cfrac{1}{\alpha_{n-1} - \cfrac{1}{\cdots\cdots,}}}$$

where the α's are independent and have the same distribution. The convergence in the particular case mentioned could have been seen directly by appealing to Worpitzky's theorem in the general theory of continued fractions, see Notes 7.5.4. For y_n we have the recursive formula

$$y_{n+1} = \cfrac{1}{\alpha_{n+1} - \cfrac{1}{y_n}},$$

which can be used to write down an integral equation for the distribution function of the y's.

An important case that leads to stochastic continued fractions is the second order difference equation

$$z_{n+1} + a_n z_n + b_n z_{n-1} = 0,$$

where the a_n, b_n are stochastic variables with distributions independent of n and such that (a_n, b_n) is independent of (a_m, b_m) if $m \neq n$. Dividing by z_n we get, putting $y_n = z_n / z_{n-1}$,

$$y_{n+1} = -a_n - \frac{b_n}{y_n},$$

which gives us a continued fraction expansion for y_{n+1}.

A more essential example is the following. Let X consist of real $k \times k$ matrices and with the usual norm $\|x\|$ of x viewed as a linear operator in R^k. Assume a probability measure defined on X such that $E\|x\| = c < 1$. Introduce the stochastic elements

$$y_n = x_n + x_n x_{n-1} + x_n x_{n-1} x_{n-2} + \dots = x_n \{ e + x_{n-1}[e + x_{n-1}(e + \dots)] + \dots \}$$

The convergence of this expression follows from the discussion in 7.3. We have clearly

$$y_{n+1} - x_{n+1} y_n = x_{n+1},$$

which is a linear first order difference equation with *stochastic coefficients* x_{n+1} and stochastic disturbances in the right member. Stochastic equations and processes of this type (taking values in a Banach algebra) can be expected to be very useful in many applications. It is easily modified to more general cases.

A problem that has been discussed a good deal (see Notes 7.5.4) is that of a random integral equation, say of the Fredholm type. Here the randomness is introduced in the integral kernel of the equation, in the given function of the inhomogeneous equation or in the domain of integration. Similarly for other stochastic linear operators. It is often possible to prove the existence of random solutions. Much more difficult seems to be the determination of the probability distribution of the solution or even its partial determination by some of its moments. This has been done so far only in very special cases and this task will no doubt attract much attention in future work.

OUTLOOK

The probability theory of algebraic structures is no doubt of practical value already in its present form and it may also look attractive to some mathematicians. The many results that are known are in the process of crystallizing into a general theory. But nothing can hide the fact that this theory is full of gaps at present, and unfortunately some of these gaps present serious difficulties when it comes to applications. Let us look at this a bit more closely.

But first let us consider some less serious (in the author's opinion) imperfections in the theory. We have not tried to present the theory in its most general form or to give necessary and sufficient conditions and so on. We have not done this even when it was possible via existing results and still less when a real effort had been necessary for this purpose. Take e.g. separability of the groups in question. One could certainly go ahead with Fourier analysis of probability distributions on a compact group without postulating the second countability axiom. And separability of the Banach spaces X and X^* may not be essential for the sort of results that we have studied. One could try to replace Borel measurability by a more general notion. Perhaps some of the statements made about stochastic groups hold without assuming local compactness, but here we may meet a barrier that is more difficult to cross. Problems of this type certainly deserve attention by such mathematicians to whom they appeal. Of course, it would be pleasant to have access to the same sort of Fourier analysis of probability distributions valid on very general algebraic-topological structures including all those that we have studied and perhaps others. It is to be feared though that such a generally applicable tool would not yield very much new and would give us only a very thin probability theory.

Even with the assumptions made there are lots of embarrasing questions left unanswered. We need only mention the form of the continuity theorem on a locally compact group or on a Banach space. Compared to what we know about the corresponding topic on the real line the situation is not satisfactory. Similarly for general homogeneous processes and infinitely divisible probability distributions: we lack a detailed characterization and a complete description. We also need a method to describe the Fourier transforms of probability distribution in the form of a criterion that works in practice.

Even if we miss complete answers to such problems we have at least partial information on these topics. We now turn to more formidable and pressing problems. On the real line we have a wealth of information about limit distributions, sometimes given implicitly and sometimes in closed form. On semi-groups, groups, etc. the situation is much worse. Take the matrix groups, some of which are of great practical interest as stochastic groups. Let us suppose that in a specified situation we have been able to show that a certain distribution can be approximated by some limiting distribution, say the normal one. How can we use this practically? Unless we happen to meet especially simple situations we do not have access to any explicit expression for this distribution. One may be able to compute the desired probability by numerical solution of the given parabolic equation. While this would be feasible we can scarcely go ahead and compute these distributions for many values of the parameters since the number of these parameters can be overwhelming, see Chapter 7. Therefore we must seek methods to express these distributions as parametric families of such basic distributions that can be computed once and for all. At present no such method is known.—It may be argued that this sort of difficulty is only what could be expected. Our knowledge of the normal distributions in finite dimensional Euclidean spaces is the result of much work, and one must be prepared to study each of the many other stochastic groups, that are of interest to us, in some detail in order to arrive at numerically useful results.

In a non-commutative structure we also run into the difficulty that the set of limit distributions (for unequal components) is not closed under convolution. Say that we study products of the form

$$u_n = s_1^{(n)} s_2^{(n)} \dots s_n^{(n)} t_1^{(n)} \dots t_n^{(n)},$$

where all the independent stochastic elements $s_\nu^{(n)}$ have the same distribution P_n and all the elements $t_\nu^{(n)}$ have the distribution Q_n. We know how to impose conditions upon P_n and Q_n in order that

$$\begin{cases} s_1^{(n)} s_2^{(n)} \dots s_n^{(n)} \to L_1, n \to \infty \\ t_1^{(n)} t_2^{(n)} \dots t_n^{(n)} \to L_2, n \to \infty. \end{cases}$$

Take the case when the limit distributions L_1 and L_2 are of the type compound Poisson. It is known that $L_1 * L_2$ need not be a compound Poisson distribution, so that the limit distribution of u_n can be of another form. This is in sharp contrast to the real line where the limit distributions often are the same for equal as for unequal component. In our more general context the case with unequal components has scarcely been explored at all and we can expect to meet real obstacles here.

Finally let us consider the following fundamental problem for non-compact stochastic groups. Sampling from the same population $P \in \mathcal{P}(G)$ we form the product $\gamma_n = g_1, g_2 \ldots g_n$ and ask for approximations to the distribution P^{n*} of γ_n. Note that the distribution P of the separate factors g_ν is kept fixed and does not tend to δ_e when the sample size increases. None of the available limit theorems can be applied directly and we must think of some new approach. The sort of result we would like to have looks as follows. There is a sequence $M_1, M_2, \ldots$ of 1—1 mappings of G onto itself and such that the elements $M_n \gamma_n$ converge distributionswise to some non-degenerate $Q \in \mathcal{P}(G)$. One is led to think of choosing the M_n's as automorphisms, exactly or approximately, and then use Fourier analysis but at present it seems doubtful that this will succeed. This is perhaps the most important of the problems left open for stochastic groups.

What we need most of all is more experience in handling particular stochastic structures. We can hope to get this when the theory is applied to special problems.

NOTES

1.2.1. We assume as known basic probability theory, e.g. Loève [1]. We will also need results from general measure theory, say to the extent of Halmos [1].

1.2.2.$_1$. One knows that a function $\varphi(z)$ is the characteristic function of a stable distribution if and only if

$$\varphi(z) = \exp\left[i\gamma z - c|z|^{\alpha}\left\{1 + i\beta\frac{z}{|z|}w(z, \alpha)\right\}\right],$$

where the constants α, β, γ, c satisfy

$$\left.\begin{array}{l} 0 < \alpha \leqslant 2 \\[4pt] -1 \leqslant \beta \leqslant 1 \\[4pt] -\infty < \gamma < \infty \\[4pt] c \geqslant 0 \end{array}\right\}$$

and
$$w(z, \alpha) = \begin{cases} \mathrm{tg}\dfrac{\pi}{2}\alpha & \text{if } \alpha \neq 1 \\[10pt] \dfrac{2}{\pi}\log|z| & \text{if } \alpha = 1. \end{cases}$$

See Gnedenko–Kolmogorov [1], p. 164.

1.2.2.$_2$. A related method to study the distribution of certain functionals of partial sums of stochastic variables is the *invariance principle* of Erdös, Kac and others, see Donsker [1]: Let $S_\nu = x_1 + x_2 + ... + x_\nu$ be the partial sum of the independent and identically distributed stochastic variables x_i with mean zero and standard deviation 1. If F is a sufficiently well behaved function of $S_1, S_2, ..., S_n$ then it can be shown that $F(S_1, S_2, ... S_n)$ has asymptotically the same distribution as $F^*(x(t))$ where F^* is a functional in $C(0,1)$, closely related to F, and where $x(t)$ is the Wiener process.

1.3.2. This problem is discussed in Grenander–Rosenblatt [1], p. 51.

1.3.3. This problem is discussed in Perrin [1], p. 1.

1.3.4. Disordered linear lattices are discussed in a great number of papers in the physics journals. Let us mention just one of outstanding interest, Dyson [1].

1.4.3. This notion of expected value is a generalization of the natural one in R^k. Say that $x = (x_1, x_2, \ldots x_k)$ is a stochastic k-vector. Then we usually define the expected value of x as the vector with the components $(Ex_1, Ex_2, \ldots Ex_n)$ if they exist. In an infinite dimensional vector space, say with a basis, there is the added difficulty that there may be no vector in the space with these given components, $Ex_\nu, \nu = 1, 2, \ldots$

1.4.7. One has also studied probability distributions in a homogeneous space. Woll [1] considered homogeneous stochastic processes taking values in such a space. Although this is related to the present subject it falls outside the framework of this book and will not be discussed here.

2.1. By separability we mean here that the space should have a countable base (second countability axiom). When we deal with metric spaces this is equivalent to the existence of a denumerable everywhere dense subset (see e.g. Dunford–Schwartz [1], p. 21); this is then the usual definition of separability for such spaces.

The separability is not always essential for what follows and at the future development of the theory the separability condition will probably be weakened a good deal. This assumption is convenient because it makes it possible to avoid certain measurability questions. Also the Hilbert spaces $L_2(G), \mathcal{H}$ etc. become separable, the set of irreducible unitary representations of a compact group becomes denumerable and so on.

It may be possible to use other σ-algebras as domain for the probability measure than those used in the text. One requirement for these σ-algebras will be that they are closed under the algebraic operations that we are using: if E, F belong to the σ-algebra then the (algebraic) product EF should also do so.

2.1.$_1$. Although not always mentioned explicitly in the text there will often be a probability space $(\Omega, \mathcal{F}, \mu)$ in the background and a stochastic element s would then be represented as $s = s(\omega)$. Here $s(\omega)$ should be a (Borel) measurable mapping: the inverse image of a Borel set in S should belong to $\mathcal{F}$.

In this connection the following problem should also be mentioned. If we want to consider an infinite system $\{s_t; t \in T\}$ of stochastic elements, is it possible to construct a probability measure on the Cartesian product S^T starting from the finite dimensional (marginal) distribution? In other words is the Kolmogorov construction (originally stated for R^T) valid in this generality?

Such questions will be discussed only occasionally in the text but we mention that this construction is possible if S is a locally compact Hausdorff space; see Pakshirajan [1]; or if S is a complete, metric space; see Špaček [2].

The regularity assumption that is made in the text is not very important and is actually not used throughout this book. Sometimes, however, it is convenient to have it, e.g. in order to have access to the Riesz representation theorem in its usual formulation. Actually, as long as we deal only with separable, locally compact spaces, every Borel measure is a Baire measure and hence regular; see Halmos [1], p. 218, 228.

The letter ω in the symbol $L_\omega(S)$ stands for the point at infinity. Given a locally compact space S we can make it into a compact space S_ω by adding a new element ω and defining the open sets of S_ω to be those of S and those that are the complements of compact subsets of S. If we also define (if S is a semi-group) $s\omega = \omega s = \omega$ then S_ω is also a semi-group, but not necessarily a topological semi-group even if S is. It goes without saying that a group loses the group property by adding ω to the group. This reduces the usefulness of this procedure (the one-point compactification) a good deal.

$2.1._2$. More details can be found in Halmos [1].

$2.1._3$. Indeed consider the P's as linear functionals of norm 1 in the space $L_\omega(S)$. Then vague convergence is the same as weak* convergence. The compactness of $\mathcal{P}(S)$ then follows from general principles, see e.g. Hille–Phillips [1], p. 37. The statement could also be verified more directly by imitating the proof on the real line and by appealing to the existence of a countable base.

$2.2._1$. See the discussion in Stromberg [1].

$2.2._2$. Suppose that $P_n \to P, Q_n \to Q$ weakly. Then for any continuous and bounded $f(s)$ we have by Fubini

$$\int_S f(s)\, P_n * Q_n(ds) = \int_S \int_S f(st)\, P_n(ds)\, Q_n(dt) = \int_S g_n(t)\, Q_n(dt),$$

where
$$g_n(t) = \int_S f(st)\, P_n(ds).$$

For any t we have $g_n(t) \to g(t) = \int_S f(st) P(ds)$. But the sequence $g_n(t)$ is equicontinuous on any compact set C. Indeed, for any $\varepsilon > 0$ we can choose a compact set C_ε such that $P_n(C_\varepsilon) \geqslant 1 - \varepsilon$. Hence

$$\left| g_n(t_2) - g_n(t_1) \right| \leqslant 2\varepsilon \sup_s \left| f(s) \right| + \max_{s \in C_\varepsilon} \left| f(st_2) - f(st_1) \right|$$

from which equicontinuity follows. This together with $g_n(t) \to g(t)$ shows, via a standard argument, that this convergence is uniform in C. Hence

$$\int_S f(s)\, P_n * Q_n(ds) - \int_S f(s)\, P * Q(ds) = \int_C \left[g_n(t) - g(t) \right] Q_n(dt)$$

$$+ \int_{C*} \left[g_n(t) - g(t) \right] Q_n(dt) + \int_S g(t)\left[Q_n(dt) - Q(dt) \right].$$

All the three integrals on the right can be made as small as desired so that $P_n * Q_n \to P * Q$ weakly.

We will not make systematic use of the fact that $\mathcal{P}(S)$ is a topological semi-group. It seems possible that this could be done and would yield useful results.

The interested reader should consult Kallianpur [1] for a general discussion of related subjects.

2.2.$_3$. The intuitive content of Theorem 2.2.2 is quite clear: the stochastic semi-group element $u = st$ runs through *exactly* $\overline{s(P) \cdot s(Q)}$, if we neglect events of probability zero. In the compact case the corresponding relation $s(P * Q) = s(P) \cdot s(Q)$ was proved by Rosenblatt [1], and for compact groups the relation has been noted by several authors.

2.2.$_4$. The following terminology would be natural. Consider a continuous homogeneous process $P_t, t \geqslant 0$, whose sample functions are almost certainly continuous. Such a process we call a Brownian motion on the semi-group. A probability distribution $\mathcal{P} \in P(S)$ is called a normal distribution if it is embeddable in a Brownian motion P_t, $P = P_1$. Since we do not study properties of sample functions this approach will not be followed up here in detail. The following properties are conjectured a) if P and Q are two normal distributions that commute then $P * Q$ is a normal distribution b) if P is a normal distribution on a locally compact group G and N is a normal subgroup of G then the induced probability distribution on the factor group $F = G/N$ is normal (see end of section 3.1).

2.3.$_1$. The reader may prefer the following slightly more intuitive definition of convergence in probability. The stochastic elements s_n with the probability distributions P_n is said to converge in probability to the constant element s_0 if $P_n(N) \to 1$ for every neighborhood N of s_0. To see that this is equivalent to weak convergence of P_n to δ_{s_0} assume that $P_n(N) \to 1$ holds. If $f(s)$ is an arbitrary continuous bounded real function on S we have

$$\int_S f(s)\,P_n(ds) = \int_N + \int_{N*} f(s)\,P_n(ds).$$

If we choose the neighborhood N of s_0 so small that $|f(s) - f(s_0)| < \varepsilon$ then the first term of the right side differs little from $f(s_0)$ and the second from zero so that

$$\int_S f(s)\,P_n(ds) \to f(s_0) = \int_S f(s)\,\delta_{s_0}(ds).$$

To prove the converse take an arbitrary neighborhood N of s_0 and choose a continuous function $f(s)$

$$\left.\begin{array}{l} 0 \leqslant f(s) \leqslant 1 \\ f(s_0) = 1 \\ f(s) = 0,\ s \notin N \end{array}\right\}$$

using e. g. Urysohn's lemma. Then

$$1 = f(s_0) \leftarrow \int_S f(s)\,P_n(ds) = \int_N f(s)\,P_n(ds) \leqslant P_n(N) \leqslant 1$$

so that $P_n(N) \to 1$.

$2.3._2$. Let us remind the reader of some basic notions from the theory of semi-groups.

A non-empty subset $L \subset S$ is called a left ideal if $SL \subset L$, and similarly a right ideal. If $I \subset S$ is both a left and right ideal it is called a twosided ideal.

An element $s \in S$ is said to be idempotent if $s \cdot s = s^2 = s$. A compact semi-group has at least one idempotent. An element $0 \in S$ is called a zero if $0s = s0 = 0$ for every $s \in S$.

We say that an idempotent s is subordinated to another idempotent t if $st = ts = s$. Further an idempotent s is called primitive if there is no non-zero idempotent subordinated to s. A semi-group S is said to be completely simple if every idempotent is primitive and to every $s \in S$ there are idempotents u and v such that $us = s = sv$. The representation $T \times X \times Y$ used later on in the text can be found in Wallace [1].

A compact semi-group has a unique minimum twosided ideal. It is denoted K and called the kernel of S. The kernel is a completely simple subsemi-group. It can be represented as

$$K = \bigcap_i SiS$$

where i runs through all the idempotents of S.

A compact semi-group for which $us = ut$ implies $s = t$ and $su = tu$ implies $s = t$ (cancellation laws) must be a group.

We shall use the relation $SIS = S$ where I is the set of all idempotent elements and we assume $SS = S$; see Koch–Wallace [1].

A very useful reference is Numakura [1] containing among other things complete proofs of the statements above.

The theorems 2.3.1, 2.3.2 a and b are due to Rosenblatt [1], [2] and to Heble-Rosenblatt [1]. Theorem 2.3.3 can be found in Glicksberg [1]. The reasoning in the proof of Theorem 2.3.2.a is due to Kloss [1].

P. Martin-Löf has suggested the following simple proof of Theorem 2.3.1. $\mathcal{P}(S)$ is compact in the weak topology. Let Q be an arbitrary point of accumulation of the sequence π_n. Then it is easy to show that $Q * P = P * Q = Q$. If Q' is some other point of accumulation it follows that $Q' = Q' * P = Q' * P^{n*} = Q' * \pi_n = Q' * Q = Q$ which proves the assertion.

If P has the Ceśaro limit π one can show that the compound Poisson process $P_t = \exp^* tP$ satisfies $P_t \to \pi$ as $t \to \infty$.

2.3.$_3$. This is done by using the equicontinuity property in a well-known way; see e.g. Dunford–Schwartz [1], p. 266.

2.3.$_4$. Here one can use a version (see Dunford–Schwartz [1], p. 662) of the mean ergodic theorem: Let T be a linear operator mapping a Banach space X into itself and such that

a) the sequence $1/n\,(T + T^2 + \ldots + T^n)\,x$ is bounded

b) for any $x \in X$ the sequence $x_n = 1/n(T + T^2 + \ldots + T^n)x$ has a subsequence converging weakly to some $\bar{x}$, and $T^n x/n$ converges to zero. Then x_n converges to $\bar{x}$. If T_0 denotes the mapping $x \to \bar{x}$, then $TT_0 = T_0 T = T_0^2 = T_0$, $\|T_0\| \leqslant C$. (Condition b) need only be assumed for x in a fundamental set.)

The operator T_0 is a projection upon the linear manifold $\{x \mid Tx = x\}$ of fixed points of T.

Theorem 2.3.1 is due to Rosenblatt [1], where the reader can also find a proof of the statement that $s(\pi)$ is the kernel of S.

2.3.$_5$. To see this one should note that for any s the set sK is an ideal, since $sKS \subset sK$, $SsK = sSK \subset sK$, and is contained in K since $sK = \bigcap_i sSiS \subset \bigcap_i SiS = K$ so that $sK = K$. This implies the existence of a unit in K and of a unique inverse so that K is a group.

Theorems 2.3.3 and 2.3.4 are due to Glicksberg [1] and Rosenblatt [1].

Let us mention a very special limit theorem. If S is a compact topological semi-group with a zero 0, $0s = s0 = 0$ for all $s \in S$, and if $0 \in s(P)$ then $P^{n*} \to \delta_0$ The proof is left to the reader.

2.3.$_6$. Throughout this book we will meet semi-groups of linear operators $T_t, t > 0$, such that $T_a T_b = T_{a+b}$ and with some sort of continuity condition. If T_t is continuous in the uniform operator topology

$$\lim_{h \downarrow 0} \|T_h - I\| = 0$$

then the infinitesimal operator

$$A = \lim_{h \downarrow 0} \frac{T_h - I}{h}$$

exists as a bounded linear transformation and $T_t = \exp tA$. If T_t is continuous in the strong topology

$$\lim_{h \downarrow 0} \|T_h x - x\| = 0, \text{ every } x \in X,$$

then the limit of $A_h = \dfrac{T_h - I}{h} x$ exists for x's forming a dense subset of X and the limit A is in general an unbounded operator. Then we can only state that

$$T_t x = \lim_{h \downarrow 0} \exp (tA_h) x.$$

For more information on this topic see Hille–Phillips [1].

2.3.$_7$. It is not easy to see what the existence of the integer m really means and how essential it is for the results. These results in the text are due to Hewitt–Zuckerman [1], where the reader can find a more complete discussion of limit results and infinite products of stochastic semi-group elements.

3.1. The basic theory of topological groups can be found in Pontrjagin [1] and Weil [1].

3.1.$_1$. The right and left invariant measures are related through $v(dg) = 1/\Delta(g)\mu(dg)$, except for an arbitrary multiplicative norming constant. Indeed for any compact set $E \subset G$ we have

$$\int_{g \in Eh} \frac{1}{\Delta(g)} \mu(dg) = \int_{u \in E} \frac{1}{\Delta(u)\,\Delta(h)} \mu(du\,h) = \int_{u \in E} \frac{1}{\Delta(u)} \mu(du)$$

so that the set function

$$\int_E \frac{1}{\Delta(g)}\, \mu\,(dg)$$

is right invariant, strictly positive and hence proportional to $\nu(E)$.

Since $\Delta(g)$ is continuous and different from zero we see that $\mu(E)=0$ is equivalent to $\nu(E)=0$ for any compact set. This implies that a probability measure which is absolutely continuous with respect to one of μ and ν is also absolutely continuous with respect to the other. We can talk of the absolutely continuous measures (with respect to the right or left invariant measures) leaving out the adjectives left and right.

3.1.$_2$. The algebra $L_1(G)$ is discussed e.g. in Hewitt [1].

3.1.$_3$. The *probability operator* T and some of its spectral properties have been discussed in Grenander [2].

The probability operators

$$T_t f(g) = \int_G f(gh)\, P_t(dh)$$

corresponding to a continuous, homogeneous process on a compact group G have some simple properties

   a) T_t is a strongly continuous semi-group in $C(G)$
   b) $T_t 1 = 1$
   c) T_t are positive, i.e. $T_t f(g) \geqslant 0$ if $f(g) \geqslant 0$ for all $g \in G$
   d) T_t commute with left translations. This is easy to verify. Let us also observe that the converse holds. Indeed if conditions a) and b) hold then for a fixed t the functional $T_t f(e)$ can be represented as

$$T_t f(e) = \int_G f(h)\, P_t(dh), \quad P_t \in \mathcal{P}(G).$$

But d) implies that for any left translation $L\ x \to gx$ we get

$$T_t f(g) = T_t Lf(e) = L\int_G f(h)\, P_t(dh) = \int_G f(gh)\, P_t(dh).$$

Finally condition a) shows that P_t forms a continuous, homogeneous process on G.

3.1.$_4$. A form of the ergodic theorem convenient for the present purpose is given in Dunford–Schwartz [1], p. 675–6.

3.1.$_5$. Some authors would use the terminology *continuous* homomorphism.

3.2. Theorem 3.2.1 is due to Wendel [1]. It is possible to make a more detailed statement about the primitive idempotents of $\mathcal{P}(G)$, see Notes

2.3.$_2$, still under the assumption that G is compact. Collins [1] has shown that P is a primitive idempotent if and only if $s(P) = H$ is a maximal proper closed subgroup of G and P is normalized Haar measure on H.

3.2.$_1$. The Fourier transform of probability distributions on compact groups was first used in the classical paper by Kawada–Itô [1] which also contains the substance of Theorem 3.2.$_4$.

Lemma 3.2.1 can be stated in a nicer form as follows.

A necessary and sufficient condition for some $\hat{P}_r, r \neq 0$, to have norm 1 is that the support $s(P)$ be contained in a closed proper subgroup A of G or a coset of A.

Proof: Suppose that $s(P) \subset g_0 A$. Then the symmetrized probability distribution $Q = \overline{P} \ast P$ has its support in A. Its Fourier transform $\hat{P}^* \cdot \hat{P}$ is Hermitian and its norm $\|\hat{P}^* \cdot \hat{P}\|$ is equal to the largest (in absolute value) eigen-value λ of $\hat{Q} = \hat{P}^* \cdot \hat{P}$. But the modulus of λ cannot be less than 1 for all the representations, $r \neq 0$, since then Q^{n*} would converge to the normed Haar measure μ on G (see Lemma 3.2.2) which has G as its support. Hence there is a z with $\|z\| = 1$ such that $\hat{P}^* \cdot \hat{P}z = \pm z$ (for some representation) and $\|\hat{P}\|$ must equal 1.

On the other hand, suppose that $\|\hat{P}\| = 1$. Then there is a z, $\|z\| = 1$, such that $\|\hat{P}z\| = 1$, i.e.

$$\left\| \int_G U(g) \, zP(dg) \right\| = 1.$$

But this implies, just as in the main text, that $s(P)$ is contained in some $g_0 A$.

Let us number all the functions $u_{ij}^{(r)}(g)$ appearing in the irreducible, unitary representations as $u_0(g) \equiv 1$, $u_1(g)$, $u_2(g)$, ... Given a set of matrices $A^{(r)}$ with the elements (with the same numbering as above) $a_0, a_1, ...$, when are they the Fourier transform of some $P \in \mathcal{P}(G)$? We give one answer to this question.

Theorem. The matrices $A^{(r)}$ form a Fourier transform of some probability distribution on the group if and only if $a_0 = 1$ and for all values of n and c_ν we have

$$\left| \sum_{\nu=0}^{n} c_\nu a_\nu \right| \leqslant \max_{g \in G} \left| \sum_{\nu=0}^{n} c_\nu u_\nu(g) \right|.$$

Proof: This is practically equivalent to a classical result of F. Riesz. The "only if" statement is obvious. The "if" part is easily proved by defining a linear functional L as follows: If

$$f(g) = \sum_{\nu=0}^{n} c_\nu u_\nu(g)$$

we define
$$Lf = \sum_{\nu=0}^{n} c_\nu a_\nu.$$

Since $|Lf| \leqslant \|f\|$, and since the $u_\nu(g)$ form a uniformly complete system in $C(G)$ it follows that L can be extended to $C(G)$ with a norm at most equal to 1 and hence there must be a completely continuous set function $P(E)$ of variation at most 1 and such that

$$Lf = \int_G f(g)\,P(dg)$$

which implies
$$Lu_\nu = a_\nu, \; A = \hat{P}.$$

Since $L1 = 1$ the set function is a probability measure.

$3.2._2$. We refer to Stromberg [1] for a modification of the Kawada–Îto statement and the related investigations by Kloss [1], [2], Urbanik [1], Hlawka [1] and Collins [2].

If P is absolutely continuous with a quadratically integrable frequency function it is possible to use the theory of Hilbert–Schmidt kernels. In this case it was shown by Rivkind [1] that P^{n*} converges to the normed Haar measure if G is connected.

$3.2._3$. This inequality is due to Kloss [2].

3.3. Theorem 3.3.2 was proved for $G = N = $ the additive group of integers by Herglotz and for $G = R^1$ by Bochner. The general case was dealt with by Raikov and Weil, see e.g. Weil [1], p. 121.

For a study of $\sup_\gamma |\hat{P}(\gamma)|$, see Hewitt–Rubin [1].

$3.3._1$. Theorem 3.3.3 and the proof of it given in the text are due to Rudin [2].

$3.3._2$. The criterion $C(P)$ is really a Hellinger integral and it may be possible to use this in order to extend the range of the definition and statement.

$3.3._3$. If the quantity $d = \sqrt{\int \|g\|^2 \mu(dg)}$, $\mu = $ normed Haar measure, is finite then it is a bound for all $d(P)$, $d(P) \leqslant d$. Indeed we have by Fubini

$$d^2 = \int_G d^2 P(dg) = \int_G \int_G \varphi(h)\,\mu(dh),$$

where
$$\varphi(h) = \int_G \|g-h\|^2 P(dg),$$

so that
$$d^2 \geqslant \inf_h \varphi(h) = d^2(P)$$

3.3.$_4$. (Added in proof). Some gaps in this context have been filled in recently both concerning infinitely divisible distributions and limit theorems. The interested reader should consult Parthasarathy–Rao–Varadhan [2] for the locally compact commutative groups and Kloss [3] for the compact commutative case. See also Carnal [1].

3.4. Finite stochastic groups were studied by Vorobev [1] and Dvoretsky–Wolfowitz [1], and the stochastic circle group by Lévy [1]. Böge [1] represented the infinitely divisible distributions on a finite group. He points out 1) that the exponential representation need not be unique, and 2) that the class of infinitely divisible distributions is closed under convolution if and only if the (finite) group is commutative.

4.1. A very useful reference to the basic theory of Lie groups is Cohn [1]; see also the later chapters in Pontrjagin [1].

4.1.$_1$. Several authors have studied processes of diffusion type on differentiable manifolds, see e.g. K. Itô [1], S. Itô [1].

4.2. The main theorem in this section is due to Hunt [1]. It generalizes the classical Lévy–Khinchin representation of infinitely divisible distributions on the real line.

The reasoning in sections 4.2–4.4 follows closely the derivations due to Hunt and Wehn.

4.2.$_1$. For a proof see e.g. Hunt [1], p. 269.

4.2.$_2$. To show that T_t maps C_k' into C_k' let us consider the case $k=1$. Then $Y'Tf = TY'f$ since T commutes with left translations. Hence T maps C_k' into C_k', and so on. Also $\|Tf\|_k' \leqslant \|f\|_k'$. See Wehn [2].

4.2.$_3$. See Wehn [1], [2]. The following related result due to Wehn is used later on. Let us consider a sequence of operators M_n of the type described in the text characterized by the constants $a_i^{(n)}$, $a_{ij}^{(n)}$ and the measures η_n, $n=1,2,\dots$. Suppose that

1) $\lim_{n\to\infty} a_i^{(n)} = a_i$

2) $\displaystyle \lim_{\varepsilon\downarrow 0} \lim_{n\to\infty} \left\{ a_{ij}^{(n)} + \frac{1}{2} \int_{0<\varphi(g)\leqslant\varepsilon} x_i(g)\,x_j(g)\,\frac{\eta_n(dg)}{\varphi(g)} \right\}$ exists and is $= a_{ij}$

3) $\lim\limits_{n\to\infty} \eta_n = \eta$ everywhere on $G_C - e$ and $\lim\limits_{n\to\infty} \eta_n(\omega) = \eta(\omega)$.

Then
$$\lim_{n\to\infty} \|(M_n - M)f\| = 0$$

for any $f \in C_2$. The proof goes as follows.

$$(M_n - M)f(g) = \sum_i (a_i^{(n)} - a_i)\, X_i f(g) + \sum_{i,j} (a_{ij}^{(n)} - a_{ij})\, X_i X_j f(g)$$

$$+ \int_{G_C - e} [f(gh) - f(g) - \sum_i X_i f(g)\, x_i(h)] \cdot \frac{\eta_n(dh) - \eta(dh)}{\varphi(h)}.$$

Decompose the integral into two parts, one where $0 < \varphi(h) \leqslant \varepsilon$, and another, where $\varphi(h) > \varepsilon$.

The norm of the second integral tends to zero because of condition 3). The first integral can again be decomposed as

$$\int\limits_{0 < \varphi(h) \leqslant \varepsilon} \left[f(gh) - f(g) - \sum_i X_i f(g)\, x_i(h) - \frac{1}{2} \sum_{i,j} X_i X_j f(g)\, x_i(h)\, x_j(h) \right]$$

$$\cdot \frac{\eta_n(dh) - \eta(dh)}{\varphi(h)} + \frac{1}{2} \sum_{i,j} X_i X_j f(g) \int\limits_{0 < \varphi(h) \leqslant \varepsilon} x_i(h)\, x_j(h) \frac{\eta_n(dh) - \eta(dh)}{\varphi(h)},$$

where the norm of the first integral tends to zero since the integrand is continuous and vanishes at the unit element e. The second integral just balances the expression.

$$\tfrac{1}{2} \sum_{i,j} (a_{ij}^{(n)} - a_{ij})\, X_i X_j f(g)$$

in the limit, see condition 2. Finally the norm of the first order terms in the original expression

$$\sum_i (a_i^{(n)} - a_i)\, X_i f(g)$$

tend to zero because of condition 1).

4.2.$_4$. Very little is known at present about normal distributions on a Lie group. We do not even know if in general the convolution of two normal distributions (corresponding to infinitesimal generators M_1 and M_2) is again normal. If M_1 and M_2 commute then the statement holds.

If P is a normal distribution on the Lie group G and N is a closed normal subgroup, then the probability distribution P' induced on the factor group $G' = G/N$ (see end of section 3.1) is also normal. To see this embed P in P_t, $P_1 = P$. Then we can define P'_t by the relation

$$\int_{G'} f(g')\, P'_t(dg') = \int_G f[h(g)]\, P_t(dg)$$

for any $f \in L(G')$ and where h is the natural homomorphism $G \to G'$. If $f \in L(G') \cap C_2(G')$ then the infinitesimal generator M' of the semi-group P'_t is given through

$$M'f(g') = Mf[h(g)],$$

where M is the infinitesimal generator of P_t. Now recall the form of M in terms of the Lie algebra Λ spanned by $X_1, X_2, \ldots X_d$. The function $f[h(g)]$ is constant on cosets of the subgroup N, and the infinitesimal transformations belonging to the Lie algebra n associated with N give zero when applied to this function. In other words the infinitesimal generator M' can be expressed as a first order plus a second order term in the infinitesimal transformations belonging to the quotient algebra Λ/n. But this is isomorphic to the Lie algebra of the group $G/N = G'$ (see Cohn [1], p. 132) and this completes the proof.—Compare this to what was said in 4.5.1 about the form of a normal frequency function expressed in that associated with a covering group.

4.2.$_5$. See e.g. Pontrjagin [1], p. 207. It is a simple consequence of the properties of the unitary representations.

4.3–4.4. Some of the derivations in these sections have only been sketched in order not to hide the mathematical substance of the reasoning. In particular some approximation arguments have only been hinted at if mentioned at all in the text. A reader wishing more detailed proofs should consult the original papers of Wehn [1], [2].

4.5.1. The construction of the process on the covering group G^* is due to S. Itô. The notion of a covering group is discussed in Cohn [1]. For Brownian motions in Lie groups and their infinitesimal operators see K. Itô [2].

In an interesting paper McKean [1] has constructed the Brownian motion by starting from a Brownian motion in the Lie algebra and forming a product integral. His construction is carried out for the three dimensional rotation group but should be possible for more general Lie groups. See also section 7.1.

The following counterexample is due to P. Martin–Löf. Let $f(x,\sigma)$ be the normal frequency function on T^1

$$f(x,\sigma) = \frac{1}{\sqrt{2\pi}\sigma} \sum_{\nu=-\infty}^{\infty} \exp - \frac{(x - 2\nu\pi)^2}{2\sigma^2}$$

and define

$$f_1(x) = f\left(x; \frac{\sigma}{\sqrt{2}}\right) + u \cos x$$

$$f_2(x) = f\left(x; \frac{\sigma}{\sqrt{2}}\right) - 2\,\frac{u\alpha}{u + 2\alpha}\, \cos x,$$

where

$$\alpha = \int_{T^1} f\left(x; \frac{\sigma}{\sqrt{2}}\right)\cos x\,dx.$$

If u is small enough then $f_1(x)$ and $f_2(x)$ are frequency functions. But

$$f_1 * f_2(x) = f\left(x; \frac{\sigma}{\sqrt{2}}\right) * f\left(x; \frac{\sigma}{\sqrt{2}}\right) + u\left(1 - \frac{2\alpha}{u + 2\alpha}\right) f\left(x; \frac{\sigma}{\sqrt{2}}\right) * \cos x$$

$$- 2\,\frac{u^2\alpha}{u + 2\alpha}\, \cos x * \cos x = f(x; \sigma)$$

since

$$\begin{cases} \cos x * \cos x = \tfrac{1}{2} \cos x \\[2mm] f\left(x; \frac{\sigma}{\sqrt{2}}\right) * \cos x = \alpha \cos x. \end{cases}$$

This shows that the normal distribution on T^1 can be factored into non-normal components.

4.5.2. See Grenander [3].

4.5.3. A more detailed discussion can be found in Karpelevich–Tutubalin–Shur [1]. The same stochastic group has been studied in Getoor [1] where the infinitely divisible distributions are examined using the Legendre transform.

5.1. A comprehensive and beautifully written introduction to locally compact groups, unitary representations, positive definite functions etc. is Neumark [1] and the interested reader should consult Godement [1], [2] for supplementary information on these topics. The direct integral representation is discussed in Mautner [1] and more completely in Mautner [2].

See also Hewitt [1], Loomis [1].

5.1.$_1$. The real meaning of the properties P and P^1 is not yet well understood. A compact group has trivially this property. A locally compact, commutative group can also be shown to have this property using a lemma in Neumark [1], p. 423. But for general, locally compact groups the answer is not known.

In the literature one has discussed the problem whether the constant function 1 can be uniformly approximated by functions of the form $c(g) * \tilde{c}(g)$.

This is related to the question of what the norm of the probability operator T is (see 3.1). Yoshizawa [1] has constructed the following example showing that this approximation is not always possible with $c \in L_2(G)$. Let G be the free group with two generators a and b. Then there is no function $c \in L_2(G)$ such that

$$\begin{cases} c * \tilde{c}(e) = 1 \\[2mm] |c * \tilde{c}(a) - 1| < \dfrac{\varepsilon^2}{2} \\[2mm] |c * \tilde{c}(b) - 1| < \dfrac{\varepsilon^2}{2} \end{cases}$$

for a given but small positive number ε. Indeed, if this were so one would have

$$\int_G |c(g) - c(hg)|^2 \mu(dg) < \varepsilon^2, \quad h = a \text{ and } b,$$

since

$$\int_G |c(g) - c(hg)|^2 \mu(dg) = 2 \int_G |c(g)|^2 \mu(dg) - 2 \operatorname{Re} c * \tilde{c}(h) < \varepsilon^2, \quad h = a \text{ and } b.$$

Consider the subset S_a of G of reduced words starting with some (non-zero) power of a and similarly S_b. Introduce the absolutely continuous probability measure

$$m(E) = \int_E |c(g)|^2 \mu(dg).$$

Then we have

$$|m(aE) - m(E)| = \left| \int_E \{ |c(ag)|^2 - |c(g)|^2 \} \mu(dg) \right| \leqslant 2\varepsilon$$

so that

$$m(aE) > m(E) - 2\varepsilon.$$

Now take E as S_a^*, aS_a^*, $a^2 S_a^*$. Then $m(S_a^*) + m(aS_a^*) + m(a^2 S_a^*) > 3m(S_a^*) - 4\varepsilon$. But $m(S_a^*) + m(aS_a^*) + m(a^2 S_a^*) \leqslant m(G) = 1$ so that

$$m(S_a^*) < \tfrac{1}{3} + \tfrac{4}{3}\varepsilon$$

and in the same way $\qquad m(S_b^*) < \tfrac{1}{3} + \tfrac{4}{3}\varepsilon.$

From this it follows, since $S_a \cap S_b = \phi$, $S_a^* \cup S_b^* = G$, that

$$1 = m(G) \leqslant m(S_a^*) + m(S_b^*) < \tfrac{2}{3} + \tfrac{8}{3}\varepsilon,$$

which cannot hold for small values of ε. This proves the impossibility of uniform approximation of 1 by $c * \tilde{c}(g)$ with c in $L_2(G)$.

Takenouchi [1] discusses P^1-groups and characterizes some of them. See also Dieudonné [1], [2], [3].

$5.1._2$. See Pontrjagin [1], p. 54.

$5.1._3$. Statement c) is due to Godement [2].

5.2. The original form of f, second part, is due to Grenander [5] who assumed that the group had the property P in order that the continuity theorem should hold. Loynes [1] showed that this assumption was not necessary. The first part of statement f) is due to Godement [1].

Theorem 5.2.1 is due to Loynes [1].

$5.2._1$. A locally compact group is complete, i.e. every Cauchy sequence has a limit; see Bourbaki [1], p. 27.—For a more general discussion of different modi of stochastic convergence in uniform spaces see Doss [2].

$5.2._2$. Pakshirajan [1] has studied convergence criteria for locally compact, commutative metric groups. He states among other things that for $\sum_1^\infty g_n$ to converge almost certainly it is necessary and sufficient that $\prod_1^\infty E\gamma(g_n)$ converges for any character $\gamma \in G^*$.

$5.2._3$. The two mentioned types of convergence are also equivalent to convergence in probability, see e.g. Loève [1], p. 251.

$5.2._4$. This is an approximation argument of standard form. See Neumark [1] p. 382 and p. 411.

$5.3._1$. For the spectral representation of a unitary operator see e.g. Riesz–Nagy [1], p. 278. This and the following integrals should actually be understood to be taken over the half-open interval $[-\pi, \pi)$.

$5.3._2$. For the concept and properties of a Bochner integral see e.g. Hille–Phillips [1], p. 78.

$5.3._3$. Such an operator-valued function as $N(\lambda)$ or $M(\lambda)$ is called a generalized resolution of the identity, Achieser–Glasmann [1], p. 263. Working with such functions it may be useful to know that they can be represented as follows. There is a Hilbert space $\mathcal{H}^+$, in which $\mathcal{H}$ is embedded, and an ordinary resolution of the identity $E^+(\lambda)$ in $\mathcal{H}^+$, such that for any $z \in \mathcal{H}$ we have $N(\lambda)z = PE^+(\lambda)z$. P is the projection operator of $\mathcal{H}^+$ on $\mathcal{H}$.

5.4. See Grenander [5].

$5.4._1$. The Stone representation of a group of unitary operators is treated e.g. in Riesz–Nagy [1], p. 380.

5.5.2. The group of linear transformations $x \to \alpha x + \beta$ and its irreducible representations are discussed in Neumark [1], p. 389. For other classical groups, see Gelfand–Neumark [1].

5.5.3. The important relation $r = \sqrt{(2h-1)/h^2}$ was proved by Kesten [1], in which paper a number of related results can be found. Let us mention the following simple expression for the spectral radius of M. Write M in its spectral representation

$$M = \int_{\lambda_1}^{\lambda_2} \lambda \, dE(\lambda),$$

where we have chosen λ_1 as large and λ_2 as small as possible, i.e. as the left and right extreme of the spectrum. Of course, the spectrum is completely determined by the diagonal elements $E_{ii}(\lambda)$, (we represent the operators $E(\lambda)$ in the l_2-space). Now P (going from i back to i in n steps) $= M_{ii}^n =$

$$= \int_{\lambda_1}^{\lambda_2} \lambda^n \, dE_{ii}(\lambda).$$

But all the M_{ii}^n and hence all the E_{ii} are independent of i as seen from their probabilistic meaning. Thus the spectral radius r is given by

$$r = \max |\lambda_1|, |\lambda_2| = \varlimsup_{n \to \infty} M_{ii}^n = [\text{radius of convergence of } \sum_n M_{ii}^n x^n]^{-1}.$$

Kesten [2] also shows the following. Let P be a symmetric probability measure on a countable group such that the support $s(P)$ spans G. Then $\|T\| = 1$ if and only if there exists a *full Banach mean value* on G. A full Banach mean value is a functional L defined on the space $B(G)$ of real valued and bounded functions on G such that

$$\begin{cases} L(c_1 f_1 + c_2 f_2) = c_1 L(f_1) + c_2 L(f_2) \\ \inf_{g \in G} f(g) \leqslant L(f) \leqslant \sup_{g \in G} f(g) \\ Lf(agb) = Lf(g), \end{cases}$$

where $f, f_1, f_2 \in B(G)$, c_1 and c_2 are arbitrary real constants and a and $b \in G$.

6.1. In this chapter we will assume some knowledge of the basic theory of linear spaces, see e.g. the first two chapters of Hille–Phillips [1].

The pioneer work in stochastic Banach spaces is Mourier [1] and contains many of the results studied in this chapter as well as much material not discussed here. Prohorov [1] is also a very good source of information especially concerning convergence of probability distributions in a Banach space as well as the two penetrating studies Prohorov [2], Le Cam [1]. See also Varadarajan [1].

6.1.$_1$. To show that $\mathcal{L}$ contains all the open subsets of X we first note that it is enough to prove that all spheres of X belong to $\mathcal{L}$. Let us then consider a sphere $S = \{x \mid \|x\| \leqslant 1\}$. Introduce the set F of all $x^* \in X^*$ such that $\|x^*\| = 1$. Then

$$S = \bigcap_{x^* \in F} A_{x^*}, \; A_{x^*} = \{x \mid |x^*(x)| \leqslant 1\},$$

since $S \supset \bigcap A_{x^*}$ is obvious, and the other inclusion relation is proved as follows: if $x \in \bigcap A_{x^*}$ and $x^* \in F$ with $x^*(x) = \|x\|$, then $\|x\| \leqslant 1$ and $x \in S$.

Now choose a sequence $x_1^*, x_2^*, \ldots$ weakly dense in F (this can be done under our separability assumption) and define

$$\left. \begin{array}{l} E_i = \{x \mid |x_i^*(x)| \leqslant 1\} \\ E = \bigcap_i E_i \end{array} \right\}$$

Obviously $E \supset S$. On the other hand, if x belongs to E but not to $S = \bigcap A_{x^*}$, then

$$\begin{cases} |x_i^*(x)| \leqslant 1, \text{ all } i \\ |x^*(x)| > 1, \text{ some } x^* \in F. \end{cases}$$

But this leads to a contradiction since we can approximate this x^* by the x_i^*, and the assertion is proved.

It was proved in Cramér–Wold [1] that a probability distribution in R^k is uniquely determined by the knowledge of the probability mass in every half space $c_1 x_1 + c_2 x_2 + \ldots + c_k x_k < c$. The half spaces form a family of unicity.

As long as we deal with separable Banach spaces we have seen that we can just as well work with the half-spaces as with the open sets; the σ-algebras generated by these two families coincide.

In the present case our probability space is not locally compact so that we do not want to define convolutions exactly as in Chapter 2. However, if X is separable and if x_1 and x_2 are two stochastic elements their sum $x_1 + x_2$ is also a stochastic element. Indeed let $x_1(\omega)$, $x_2(\omega)$ be the two stochastic elements defined on a probability space $(\Omega, \mathcal{A}, P)$. From what we have just seen it is sufficient if we can show that sets of the form

$$\{\omega \mid x^*[x_1(\omega) + x_2(\omega)] \leqslant 1\} = \{\omega \mid x^*[x_1(\omega)] + x^*[x_2(\omega)] \leqslant 1\}$$

belong to the σ-algebra $\mathcal{A}$. But $x^*[x_1(\omega)]$ and $x^*[x_2(\omega)]$ are ordinary real valued stochastic variables so that this is true for every $x^* \in X^*$. Thus the stochastic element $x_1(\omega) + x_2(\omega)$ has well-defined probabilities on the Borel sets of X. This no longer holds in a general not separable space X; an example can be found in Nedoma [1].

The following criterion is sometimes useful. Let $x_n(\omega)$ be a sequence of stochastic elements in X such that $x_n(\omega) \to x(\omega) \in X$ for every ω. Then $x(\omega)$ is a stochastic element. For more information on this topic see Hanš [1].

The convolution $P * Q$ is continuous in P, Q. Indeed assume that $P_n \to P$, $Q_n \to Q$ weakly. Then for any bounded and continuous $f(x)$ we have with $R_n = P_n * Q_n$

$$\int_X f(x)\, R_n(dx) = \int_{x \in X} P_n(dx) \int_{y \in X} f(x+y)\, Q_n(dy)$$

$$\to \int_{x \in X} P(dx) \int_{y \in X} f(x+y)\, Q(dy) = \int_X f(x)\, R(dx)$$

by a modification of the proof in Notes 2.2.$_2$, see also Theorem 6.2.2.

6.1.$_2$. In the finite dimensional Euclidean space this was shown already in Cramér–Wold [1].

6.1.$_3$. Proof of the statements 1)–5) of the mean value operation (in terms of the Pettis integral) can be found e.g. in Hille–Phillips [1], pp. 77–85.

It may be remarked that for the separable spaces that we consider the Pettis integral is equivalent to the Bochner integral and $E\|x\| < \infty$ is not only a sufficient but also a necessary condition for Ex to exist.

Some authors have introduced a mathematical expectation or mean value in more general spaces. If X is a metric space with the metric $\varrho(x,y)$ then a stochastic element x taking values in X is said to have a mean value $m \in X$ if

$$E\varrho^2(x, m) = \inf_{a \in X} E\varrho^2(x, a).$$

This definition is due to Fréchet [1]. A similar one due to Doss [1] is the following: The stochastic element x has the mean value m in X if

$$\varrho(m, a) \leqslant E\varrho(x, a)$$

for every $a \in X$. Since these definitions fall outside the framework of this book they will not be studied further here; the interested reader should consult the two references mentioned above as well as Urbanik [3] for a study of the Doss integral in case X is a commutative group with the invariant metric ϱ.

6.1.$_4$. This metric, that had been introduced on the real line by Lévy, was studied in the present context by Prohorov [1], where proofs can be found of the statement concerning the metrization.

6.2.$_1$. See Halmos [1], p. 40. In a non-separable space the situation has not been explored completely. See also Driml [1].

6.2.$_2$. See Prohorov [1]. On the real line the compactness of the probability measures P_n is equivalent to the equicontinuity of $\hat{P}_n$ at $x^*=0$. It deserves to be mentioned that there is no locally convex topology in the Hilbert space l_2 for which this statement holds. See Prohorov–Sazonov [1].

6.2.$_3$. Theorem 6.2.4 is due to Kolmogorov [3]. Some interesting extensions are given in Prohorov [2].

It has been suggested that one should work with weak distributions: if $x^*(x)$ have well-defined and consistently given probability distributions for all $x^* \in X$ then we call this system of probability distributions a weak distribution over X. It is not difficult to see that $\varphi(x^*)$ defines a weak distribution if and only if it is positive definite, $\varphi(0)=1$ and $\varphi(x^*)$ is continuous along any ray $c \cdot x^*$, $-\infty < c < \infty$. At present the usefulness of the weak distributions seems questionable.

6.3. The normal distributions could also have been defined equivalently as follows. If $\mathcal{P} \in P(X)$ we say that it is a normal probability distribution if $x^*(x)$ is a normal distribution (on R^1) for every $x^* \in X^*$.

6.3.$_1$. This reasoning is based upon the following theorem of Cramér [1]: If x, y, z are real-valued stochastic variables, x and y are independent and $x = x + y$ is normal, then x and y also have normal distributions.

6.3.$_2$. In the proof of Theorem 6.3.6 one need only assume that

$$\lim_{N \to \infty} \sup_n \sum_{N+1}^{\infty} s_\nu^n = 0;$$

actually this is also a necessary condition. See Prohorov [1].

6.4. There are many possible versions of the law of large numbers in a Banach space, several of them due to Mourier [1]. In some of them independence is replaced by stationarity. Beck [1] has considered a sequence of independent stochastic elements x_n with $Ex_n = 0$, $E\|x_n\|^2 \leqslant M < \infty$, and shown that then a strong law of large numbers holds under the assumption that X is a uniformly convex Banach space.

It is possible and not difficult to replace the assumption that the terms x_ν be independent and identically distributed by the condition that they form a stationary sequence. Similarly one can study continuous parameter stationary processes with values in a Banach space and the corresponding ergodic theorems.

6.5. These central limit theorems are due to Mourier [1] and Fortet–Mourier [2]. X is said to be *G-space* if there is a constant K and a mapping $g:X\to X^*$, $g(x)\in X^*$ such that

$$1)\ \ \|g(x)\| = \|x\|$$
$$2)\ \ (g(x),x) = \|x\|^2$$
$$3)\ \ \|g(x)-g(y)\| \leqslant K\,\|x-y\|$$

Among the G-spaces are the L_p-spaces with $p\geqslant 2$ and in particular the Hilbert space.

6.6. Stochastic Schwartz distributions were introduced by Gelfand [1] and K. Itô [3]. Further work has been done by Ullrich [1], Urbanik [3] and others.

Gelfand [1] uses the following somewhat wider definition of a stochastic Schwartz distribution: $d(\varphi)$ should be an ordinary stochastic variable for any φ and similarly for any vector $d(\varphi_1),d(\varphi_2),\dots d(\varphi_n)$; d should be a linear functional $d(c_1\varphi_1+c_2\varphi_2)=c_1d(\varphi_1)+c_2d(\varphi_2)$; for any natural number n and $\varphi_{i\nu}$ such that $\lim\limits_{\nu\to\infty}\varphi_{i\nu}=\varphi_i$; $i=1,2,\dots n$; the (finite-dimensional) distribution of $d(\varphi_{1\nu})$, $d(\varphi_{2\nu}),\dots d(\varphi_{n\nu})$ should tend to that of $d(\varphi_1),d(\varphi_2),\dots,d(\varphi_n)$. He introduces the characteristic functional

$$\hat{P}(\varphi) = E\,\exp id(\varphi),\ \varphi\in\Phi.$$

As usual this a Hermitian positive definite function with $\hat{P}(0)=1$. It is continuous since if $\varphi_n\to\varphi$ then the distributions of the ordinary stochastic variables $d(\varphi_n)$ tend to that of $d(\varphi)$ so that $\hat{P}(\varphi_n)\to\hat{P}(\varphi)$. — What is more interesting is that the converse holds. First it is clear that if $f(\varphi)$ is positive definite, Hermitian and $f(0)=1$, then we can define consistently finite-dimensional probability distributions of (ordinary) stochastic variables that we denote by $x_1=d(\varphi_1)$, $x_2=d(\varphi_2),\dots x_n=d(\varphi_n)$, by demanding that

$$E\,\exp i\sum_1^n t_\nu x_\nu = f\Big(\sum_1^n t_\nu\varphi_\nu\Big).$$

Let $x=d\Big(\sum_1^n c_\nu\varphi_\nu\Big)$ be the stochastic variable corresponding to a linear combination $\sum_1^n c_\nu\varphi_\nu$. Then

$$E\,\exp iu\Big(x-\sum_1^n c_\nu x_\nu\Big)\equiv f(0)=1$$

so that $P(x - \sum_1^n c_\nu x_\nu = 0) = 1$ and $d(\sum_1^n c_\nu \varphi_\nu) = \sum_1^n c_\nu d(\varphi_\nu)$ almost certainly as desired. Finally if $\{\varphi_{i\nu}\}$ is as above then the distributions just introduced for the stochastic variables $d(\varphi_{1\nu})$, $d(\varphi_{2\nu})$, $\ldots d(\varphi_{n\nu})$ converge to those of $d(\varphi_1)$, $d(\varphi_2)$, $\ldots d(\varphi_n)$ as ν tends to infinity. Hence the positive definite function $f(\varphi)$ corresponds to a stochastic Schwartz distribution as defined by Gelfand.

Compare this notion to that of a weak distribution in a Banach space, Notes 6.2.3.

6.6.1. Theorem 6.6.1 is due to Ullrich who proves it by an approximation argument: there is a sequence $\varphi_1, \varphi_2, \ldots$ in Φ with the following property. For any $\varphi \in \Phi$ it can be approximated by a subsequence $\varphi_{n_1}, \varphi_{n_2}, \ldots$ But the convergence of

$$\frac{1}{n} \sum_1^n d_\nu(\varphi_i)$$

is a direct consequence of the classical law of large numbers. Using the equicontinuity it can be seen that

$$\frac{1}{n} \sum_1^n d_\nu(\varphi) \to m(\varphi)$$

for all $\varphi \in \Phi$ with probability one.

6.7. A more complete discussion of this topic can be found in Fortet–Mourier [1]. They study the convergence of the empirical distribution functions (certain random probability measures) on a general metric space. Their results are not direct consequences of the previous version of the law of large numbers but require a separate derivation.

Another application is to Orlicz spaces. Let $v = \varphi(u)$ and $u = \psi(v)$ be two functions inverse to each other and such that 1) $\varphi(0) = \psi(0) = 0$ and 2) φ and ψ are non-decreasing. Introduce the functions

$$\Phi(u) = \int_0^u \varphi(x)\, dx, \quad \Psi(v) = \int_0^v \psi(x)\, dx.$$

On a measure space $(A, \mathcal{A}, m)$ we consider the $\mathcal{A}$-measurable complex valued functions $f(a)$ such that

$$\sup \int_A |f(a) g(a)|\, m(da) < \infty,$$

with the supremum taken over all g such that

$$\int_A \Psi[|g(a)|]\, m(da) \leqslant 1.$$

This set of functions L_Φ is called an Orlicz space. It can be made into a Banach space (see Zaanen [1], p. 101). In a similar way L_Ψ is introduced. They generalize the L_p spaces. Now we can consider stochastic elements taking values in L_Φ and apply the general results of Chapter 6. See Bharucha-Reid [1], [2].

7.1.$_1$. To see that $P_x \circ P_y$ is well-defined note that the stochastic elements $x(\omega)$ and $y(\omega)$ can be approximated (strongly) by stochastic elements $x_n(\omega)$ and $y_n(\omega)$ that take only denumerably many values. The product $x_n(\omega) \cdot y_n(\omega)$ defines a probability measure (it is also denumerably valued). Since the product operation is continuous we have $x_n(\omega) \cdot y_n(\omega) \to x(\omega) \cdot y(\omega)$ and it follows (see Notes 6.1.$_1$) that $x(\omega) \cdot y(\omega)$ is a stochastic element with a well-defined probability measure. We have furthermore

$$\int_{X \times X} f(xy)\, P_x(dx)\, P_y(dy) = \int_X f(z)\, R(dz)$$

with $R = P_x \circ P_y$ for any bounded and continuous function $f(x)$ on X.— Alternatively we could start from the relation above.

   Suppose that $P_n \to P, Q_n \to Q$ weakly. Then it can be seen, as before, that $P_n \circ Q_n \to P \circ Q$ weakly.

7.1.$_2$. This expression may remind the reader of the concept of a multiplicative integral. There is an abundance of definitions of such integrals in the literature, but none of them seems to be suitable for the present purpose.

7.1.$_3$. It is of some interest to observe that Theorem 7.1.5 is a special case of Theorem 7.1.3. Indeed condition C_1' is satisfied here since

$$\sum_{\nu=1}^{n} E\|y_{n\nu}\| = \frac{1}{n} \sum_{\nu=1}^{n} E\|y_\nu\| = E\|y_1\| < \infty.$$

Further for any $c \in (0,1)$ we have almost certainly

$$\sum_{\nu=1}^{[cn]} y_{n\nu} = \frac{1}{n} \sum_{\nu=1}^{[cn]} y_\nu \xrightarrow[\text{strongly}]{} c \cdot Ey_1,$$

which follows from the law of large numbers in the Banach space. But then Theorem 7.1.3 tells us that the product γ_n should tend in probability to the stochastic element $\exp m$, which here reduces to a constant element.

7.2.$_1$. For this and related statements in the text see Hille–Phillips [1], pp. 118 f.

7.2.$_2$. In the matrix case this result is due to Furstenberg–Kesten [1] in which case these authors also showed that $1/n \log\|x_1 x_2 \dots \gamma_n\|$ actually converges almost certainly.

$7.2._3$. See Hille–Phillips [1], p. 124.

$7.2._4$. See e.g. Halmos [1], p. 113.

7.3. The regularity properties of stochastic operators, resolvents, etc. have been studied by several authors among whom we mention Bharucha-Reid [1], [2], Hanš [2], Špaček [1]. Theorem 7.3.1 is due to Hanš.

$7.3._1$. See Banach [1].

$7.3._2$. The idea of the random ergodic theorems goes back to Ulam and von Neumann; the version of it stated in Theorem 7.3.3 is due to Kakutani [1]. An interesting application of this is the following, see Robbins [1]. Let G be a compact group and let the random transformations x_ω be defined as gh of G onto G, where h has a probability distribution over G. Then x_ω is measure preserving with respect to Haar measure and it follows from the random ergodic theorem that

$$\lim_{n \to \infty} \frac{1}{n} \sum_{\nu=1}^{n} f(h_1 h_2 \ldots h_\nu)$$

exists almost certainly. Further it can be shown in this particular case that the limit is a constant almost certainly.

$7.4._1$. See Neumark [1], p. 189.

$7.4._2$. See Neumark [1], p. 210.

7.5. Bellman [1] used Kronecker's products to study products of stochastic matrices, especially to obtain the moments of the entries of the product matrix.

7.5.1. The treatment of the matrix valued multiplicative processes is due to Grenander [3].

7.5.2. The statements of this section have been prooved in Furstenberg–Kesten [1].

7.5.3. See Anderson [1], Chapter 13, and Nyquist–Rice–Riordan [1].

The asymptotic spectrum of the stochastic, symmetric matrices was obtained by Wigner [1]. We would like to add the following observation. To show directly the convergence of the strength function mentioned in the text one could proceed as follows. Introduce the characteristic functions

$$\varphi_n(z) = \int_{-\infty}^{\infty} e^{i\lambda z} dF_n(\lambda) = \frac{1}{n} \sum_{\nu=1}^{n} e^{i\lambda_\nu z}.$$

We have
$$\varphi_n(z) = \sum_{p=0}^{2k-1} \mu_p^{(n)} \frac{(iz)^p}{p!} + 0\left(\mu_{2k}^{(n)} \frac{z^{2k}}{(2k)!}\right)$$

so that we have when $n \to \infty$

$$E\varphi_n(z) \to \sum_{p=0}^{2k-1} c_{2p} \frac{(iz)^p}{p!} + 0\left(c_{2k} \frac{z^{2k}}{(2k)!}\right).$$

The error term can be made as small as desired for fixed z by choosing k large. Hence

$$E\varphi_n(z) \to \varphi(z) = \frac{1}{\pi} \int_{-2}^{2} e^{i\lambda z} \sqrt{1 - \frac{u^2}{4}}\, du$$

for any real value of z. But introducing $EF_n(\lambda) = G_n(\lambda)$ this is equivalent to

$$\int_{-\infty}^{\infty} e^{iz\lambda} dG_n(\lambda) \to \varphi(z)$$

which is possible only if

$$G_n(\beta) - G_n(\alpha) \to \frac{1}{\pi} \int_{\alpha}^{\beta} \sqrt{1 - \frac{u^2}{4}}\, du,$$

so that the distributions G_n tend to a limiting distribution with the strength function

$$s(u) = \frac{1}{\pi} \sqrt{1 - \frac{u^2}{4}}.$$

While the results discussed in section 7.5.3 have a good deal of independent interest, one has the feeling that a general theory of spectra of stochastic operators must use more of the essential algebraic properties of the operators. (Added in proof). See also F. J. Dyson: A Brownian-motion model for the eigen-values of a random matrix, J. Math. Phys., Vol. 3, 1962.

7.5.4. See Wall [1], p. 42. For a study of stochastic integral equations see Bharucha-Reid [1], [2].

BIBLIOGRAPHY

ACHIESER, N. J.–GLASMANN, J. M. [1]: Theorie der linearen Operatoren im Hilbert-Raum, Berlin, 1954.

ANDERSON, T. W. [1]: An introduction to multivariate statistical analysis, New York, 1958.

BANACH, S. [1]: Théorie des opérations linéaires, Warszawa, 1932.

BECK, A. [1]: Une loi forte des grands nombres dans des espace de Banach uniformément convexes, Ann. Inst. Henri Poincaré, XVI, 1958.

BELLMAN, R. [1]: Limit theorems for non-commutative operations, I, Duke Math. J., 21, 1954.

BERGSTRÖM, H. [1]: Konvergenzsätze über unendliche Produkte in abstrakten Halbgruppen, Abh. Math. Semin. Hamburg, 1959.

BHARUCHA-REID, A. T. [1]: On random elements in Orlicz spaces, Bull. Acad. Pol. Sci., IV, 1956.

—— [2]: Über die Konvergenz der Folgen von verallgemeinerten zufälligen Grössen in Orlicz'schen Räumen, Bull. Acad. Pol. Sci., VII, 1959.

—— [3]: On random solutions of Fredholms integral equations, Bull. Am. Math. Soc., 66, 1960.

—— [4]: On random solutions of integral equations in Banach spaces, Trans. Second Prague Conf., Inform. theory. Stat. decision functions. Random Processes, Prague 1960.

BOCHNER, S. [1]: Harmonic analysis and the theory of probability. Berkeley 1955.

BÖGE, W. [1]: Über die Charakterisierung unendlich teilbarer Wahrscheinlichkeitsverteilungen, J. reine und angew. Math., 201, 1959.

BOURBAKI, N. [1]: Topologie Générale (Livre III, chapitre III), Paris 1951.

CARNAL, H.: Distributions aléatoires indéfiniment divisible sur les groupes compacts, C.R., Paris, t. 255, 1962.

CIGLER, J. [1]: Folgen normierter Masse auf kompakten Gruppen, Zeits. Wahrscheinlichkeitstheorie u. verw. Gebiete, 1, 1962.

COHN, P. M. [1]: Lie groups, Cambridge 1957.

COLLINS, H. S. [1]: Primitive idempotents in the semigroup of measures, Duke Math. J., 27, 1960.

—— [2]: Convergence of convolution iterates of measures, Duke Math. J., 29, 1962.

CRAMÉR, H. [1]: Über eine Eigenschaft der normalen Verteilungsfunktion, Math. Z., 41, 1936.

—— [2]: Mathematical methods of statistics, Princeton 1946.

CRAMÉR, H.–WOLD, H. [1]: Some theorems on distribution functions, J. London Math. Soc., 11, 1936.

DIEUDONNÉ, J. [1]: Sur le produit de composition, Comp. Math., 12, 1954.

—— [2]: Sur le produit de composition II, J. Math. P. Appl. 39, 1960.

—— [3]: Sur une propriété des groupes libres, J. R. u. Ang. Math., 204, 1960.

DONSKER, M. D. [1]: An invariance principle for certain probability limit theorems, Memoirs of the Am. Math. Soc., Number 6, 1951.

DOOB, J. L. [1]: Stochastic processes, New York, 1954.

DOSS, S. [1]: Sur la moyenne d'un élément aléatoire dans un espace distancié, Bull. Sci. Math., 73, 1949.

—— [2]: Sur la convergence stochastique dans les espaces uniformes, Ann. Éc. Norm., LXXI, 1954.

DRIML, M. [1]: Convergence of compact measures on metric spaces, Trans. Second Prague Conference. Inform. theory. Stat. decision functions, Random processes, Prague 1960.

DUNFORD, N.–SCHWARTZ, J. T. [1]: Linear operators. Part I: General theory. New York, 1958.

DVORETZKY, A.–WOLFOWITZ, J. [1]: Sums of random integers reduced modulo m, Duke Math. J., 18, 1951.

DYSON, F. J. [1]: The dynamics of a disordered linear chain, Phys. Rev., 101, 1956.

FORTET, R. [1]: Normalverteilte Zufallselemente in Banachschen Räumen, Anwendungen auf zufällige Funktionen, Tagung Wahrscheinl. Math. Stat. Berlin 1954.

FORTET, R.–MOURIER, E. [1]: Convergence de la répartition empirique vers la répartition théorique, Ann. É. Norm. Sup., 60, 1953.

—— [2]: Les fonctions aléatoires comme éléments aléatoires dans les espace de Banach, Stud. Math., 15, 1955.

FRÉCHET, M. [1]: Les éléments aléatoires de nature quelconque dans un espace distancié, Ann. Inst. Henri Poincaré, 10, 1947.

—— [2]: Abstrakte Zufallselemente, Tagung Wahrscheinl. Math. Stat. Berlin, 1954.

FURSTENBERG, H.–KESTEN, H. [1]: Products of random matrices, Ann. Math. Stat., 31, 1960.

GELFAND, I. M. [1]: General random processes (Russian), Dokl. Akad. Nauk SSSR, 100, 1955.

GELFAND, I. M.–NEUMARK, M. A. [1]: Unitäre Darstellungen der klassischen Gruppen, Berlin 1957.

GERTSENSHTEIN, M. E.–VASILEV, V. B. [1]: Waveguides with random inhomogeneities and Brownian motion in the Lobachevsky plane (Russian), Teoriya Veroyat. i ee Prim. IV, 1959.

GETOOR, R. K. [1]: Infinitely divisible probabilities on the hyperbolic plane, Pac. J. Math., 11, 1961.

GLICKSBERG, I. [1]: Convolution semigroups of measures, Pac. J. Math., 9, 1959.

GNEDENKO, B. V.–KOLMOGOROV, A. N. [1]: Limit distributions for sums of independent random variables, Cambridge, Mass., 1954.

GODEMENT, R. [1]: Les fonctions de type positif et la théorie des groupes, Trans. Am. Math. Soc., 63, 1948.

—— [2]: Sur la théorie des représentations unitaires, Ann. Math., 53, 1951.

GORMAN, C. D. [1]: Brownian motion of rotation, Trans. Am. Math. Soc., 94, 1960.

GRENANDER, U. [1]: Some non-linear problems in probability theory, Harald Cramér volume, New York, 1959.

—— [2]: Stochastic groups. Background. General discussion. Remarks on limit theorems. Ark. Mat. 4, 1960.

—— [3]: Stochastic groups. A particular stochastic group. Some stochastic algebras. Ark. Mat., 4, 1960.

—— [4]: Stochastic groups. Fourier analysis of probability distributions on locally compact groups. Ark. Mat., 5, 1961.

—— [5]: Stochastic groups and related structures, Fourth Berkeley Symp. on Math. Stat. and Prob., Berkeley, 1961.

GRENANDER, U.–ROSENBLATT, M. [1]: Statistical analysis of stationary time series, New York, 1957.

HALMOS, P. [1]: Measure theory, New York, 1950.

HANŠ, O. [1]: Generalized random variables. Trans. First Prague Conf. Information theory, stat. decision functions, random processes. Prague 1957.

—— [2]: Random fixed point theorems. Trans. First Prague Conf. Inform. theory. Stat. decision functions, Random processes, Prague, 1957.

HEBLE, M.–ROSENBLATT, M. [1]: Idempotent measures on a compact topological semigroup, mimeographed report.

HEWITT, E. [1]: A survey of abstract harmonic analysis, Surveys in applied math. IV, New York, 1958.

HEWITT, E.–RUBIN, H. [1]: The maximum value of a Fourier-Stieltjes transform, Math. Scand., 3, 1955.

HEWITT, E.–ZUCKERMAN, H. S. [1]: Arithmetic and limit theorems for a class of random variables, Duke Math. J., 22, 1955.

HILLE, E.–PHILLIPS, R. S. [1]: Functional analysis and semigroups, Providence 1957.

HLAWKA, E. [1]: Statistik auf kompakten Gruppen, I. Österr. Akad. Wiss., 1959.

HUNT, G. A. [1]: Semi-groups of measures on Lie groups, Trans. Amer. math. Soc., 81, 1956.

ITÔ, K. [1]: Stochastic differential equations in a differentiable manifold, Nagoya Math. J., 1, 1950.

—— [2]: Brownian motions in a Lie group, Proc. Jap. Acad., 26, 1950.

—— [3]: Stationary random distributions, Mem. Coll. Sci., Univ. of Kyoto, 28, 1954.

ITÔ, S. [1]: Fundamental solutions of parabolic differential equations and boundary value problems, Japanese J. Math., XXVII, 1957.

—— [2]: Brownian motions in a topological group and its covering group, Rend. Circ. Mat. Palermo, 1, 1952.

KAKEHASHI, T. [1]: Stationary periodic distributions, J. Osaka Inst. Sci. Technology, 1, 1949.

KAKUTANI, S. [1]: Random ergodic theorems and Markov processes with a stable distribution, Proc. Second Berkeley Symp., Math. Stat. Prob., Berkeley, 1951.

KALLIANPUR, G. B. [1]: The topology of weak convergence of probability measures, J. Math. Mech., 10, 1961.

KARPELEVICH, F. I.–TUTUBALIN, V. N.–SHUR, M. G. [1]: Limit theorems for the compositions of distributions in the Lobachevsky plane (Russian), Teoriya Veroyat. i ee Prim., IV, 1959.

KAWADA, Y.–ITÔ, K. [1]: On the probability distribution on a compact group I. Proc. Phys.-Mat. Soc. Japan, 22, 1940.

KESTEN, H. [1]: Symmetric random walk on groups, Trans. Am. Math. Soc.

—— [2]: Full Banach mean values on countable groups, Math. Scand., 7, 1959.

KLOSS, B. M. [1]: Probability distributions on bicompact topological groups, (Russian), Teoriya Veroyat. i ee Prim., IV, 1959.

—— [2]: Limit distributions for sums of independent variables with values in a bicompact group, (Russian), Dokl. Akad. Nauk SSSR, 109, 1953.

—— [3]: Limiting distributions on bicompact abelian groups, (Russian), Teoriya Veroyat. i ee Prim. VI, 1961.

KHINCHIN, A. YA. [1]: Asymptotische Gesetze der Wahrscheinlichkeitsrechnung, Ergebnisse der Math., Berlin 1933.

KOCH, R. J.–WALLACE, A. D. [1]: Maximal ideals in compact semigroups, Duke Math. J., 21, 1954.

KOLMOGOROV, A. N. [1]: Grundbegriffe der Wahrscheinlichkeitsrechnung, Ergebnisse der Math., Berlin 1933.

—— [2]: La transformation de Laplace dans les espaces linéaires, Compt. Rend., 200, 1935.

—— [3]: A note on the papers of R. A. Minlos and V. Sazonov (Russian), Teoriya Veroyat. i ee Prim., IV, 1959 (Summary of papers).

LE CAM, L. [1]: Un instrument d'étude des fonctions aléatoires: la fonctionelle caractéristique, Compt. Rend. 224, 1947.

—— [2]: Convergence in distribution of stochastic processes, Publ. Statistics, Univ. California Press, Berkeley, 1957.

LÉVY, P. [1]: L'addition des variables aléatoires définies sur une circonférence, Bull. Math. France, 67, 1939.

LOÈVE, M. [1]: Probability theory, Second edition, Princeton, 1960.

LOOMIS, L. H. [1]: An introduction to abstract harmonic analysis, New York, 1953.

LOYNES, R. [1]: Fourier analysis of probability distributions on locally compact groups, to appear.

MAUTNER, F. I. [1]: Unitary representations of locally compact groups II, Ann. Math., 52, 1950.

—— [2]: Note on the Fourier inversion formula on groups, Trans. Am. Math. Soc., 78, 1955.

MCKEAN, H. P. [1]: Brownian motions on the 3-dimensional rotation group, Mem. Coll. Sci. Univ. Kyoto, XXXIII, 1960.

MOURIER, E. [1]: Éléments aléatoires dans un espace de Banach, Ann. Inst. Henri Poincaré, 13, 1953.

NEDOMA, J. [1]: Note on generalized random variables, Trans. First Prague Conf. Inform. Theory. Stat. dec. functions. Random processes, Prague, 1957.

NEUMARK, M. A. [1]: Normierte Algebren, Berlin 1959.

NUMAKURA, K. [1]: On bicompact semigroups, Math. J. Okayama Univ., 1, 1952.

NYQUIST, H.–RICE, S. O.–RIORDAN, J. [1]: The distribution of random determinants, Quart. Appl. Math., 12, 1954.

PARTHASARATHY, K. R.–RANGA RAO, R.–VARADHAN, S. R. S. [1]: On the category of indecomposable distributions on topological groups, Trans. Amer. math. Soc., 102, 1962.

—— [2]: Probability distributions on locally compact abelian groups (to appear).

PAKSHIRAJAN, R. P. [1]: Regular measures and stochastic processes in topological groups, to appear.

—— [2]: A property of regular measures in locally compact Hausdorff spaces, Proc. Am. Math. Soc., 12, 1961.

PERRIN, F. [1]: Étude mathématique du mouvement Brownien de rotation, Ann. Sci. École Norm. Sup., 1928.

PONTRJAGIN, L. [1]: Topological groups. Princeton 1946.

PRÉKOPA, A.–RÉNYI, A.–URBANIK, K. [1]: On the limiting distribution of sums of independent random variables in commutative bicompact topological groups (Russian), Acta Math. Acad. Sci. Hung., 7, 1956.

PROHOROV, YU. V. [1]: Convergence of random processes and limit theorems in probability theory (Russian), Teoriya Veroyat. i ee Prim., 1, 1956.

—— [2]: The method of characteristic functionals, Proc. fourth Berkeley Symp. Math. Stat. Prob. Berkeley 1961.

PROHOROV, YU. V.–SAZONOV, V. V. [1]: Some results associated with Bochner's theorem, (Russian), Teoriya Veroyat. i ee Prim., 6, 1961.

RIESZ, F.–NAGY, B. SZ. [1]: Leçons d'analyse fonctionelle, Budapest, 1952.

RIVKIND, YA. I. [1]: Limit theorems of probalility theory on compact topological groups, (1955). English translation in Selected Translations in Math. Stat. Prob., Vol. 2, Providence, 1962.

ROBBINS, H. [1]: On the equidistribution of sums of independent random variables, Proc. Am. Math. Soc. 4, 1953.

ROSENBLATT, M. [1]: Limits of convolution sequences of measures on a compact topological semigroup, J. Math. Mech., 9, 1960.

—— [2]: Idempotent measures on a compact topological semigroup (to appear).

RUDIN, W. [1]: Measure algebras on Abelian groups, Bull. Am. Math. Soc., 65, 1959.

—— [2]: Idempotent measures on Abelian groups, Pac. J. Math.

SCHWARZ, Š. [1]: On the structure of the semigroup of measures on a finite semigroup, Czechosl. Math. J., 7, 1957.

ŠPAČEK, A. [1]: Prolongement des transformations aléatoires, Proc. First Prague Conf. Inf. theory. Stat. decision functions, random processes, Prague, 1960.

—— [2]: Probability measures in infinite Cartesian products, Ill. J. Math., 4, 1960.

Stromberg, K. [1]: Probabilities on a compact group, Trans. Am. Math. Soc., 94, 1960.

—— [2]: A note on the convolution of semi-groups, Math. Scand., 7, 1959.

TAKENOUCHI, O. [1]: Sur une classe de fonctions continues de type positif sur un groupe localement compact, Math. J. Okayama Univ. 4, 1955.

Ullrich, M. [1]: Some theorems on random Schwartz distributions, First Prague Conf. Information theory, stat. decision functions, random processes. Prague, 1957.

Urbanik, K. [1]: On the limiting probability distribution on a compact topological group, Fund. Math., 44, 1957.

——— [2]: Poisson distributions on compact Abelian topological groups, Colloq. Math., VI, 1958.

——— [3]: Remarks on the Doss integral, Colloq. Math., 5, 1957.

——— [4]: Generalized stochastic processes with independent values, Proc. fourth Berkeley Symp. Math. Stat. Prob. Berkeley 1961.

Varadarajan, V. S. [1]: Weak convergence of measures on separable metric spaces, Sankhyā, 19, 1958.

Vorobev, N. N. [1]: The addition of independent random variables on finite groups (Russian), Mat. Sbornik, 34, 1954.

Wall, H. S. [1]: Analytic theory of continued fractions, New York 1948.

Wallace, A. D. [1]: The Rees–Suschkewitsch structure theorem for compact simple semigroups, Proc. Nat. Acad. Sci., U.S.A., 42, 1956.

Wehn, D. F. [1]: Probabilities on Lie groups, Proc. Nat. Acad. Sci., 1962.

——— [2]: Limit distributions on Lie groups, to appear.

Weil, A. [1]: L'intégration dans les groupes topologiques et ses applications, Paris, 1940.

Wendel, J. G. [1]: Haar measure and the semi-group of measures on a compact group, Proc. Amer. Math. Soc., 5, 1954.

Wigner, E. P. [1]: On the distribution of the roots of certain symmetric matrices, Ann. Math., 67, 1958.

Woll, J. W. [1]: Homogeneous stochastic processes, Pac. J. Math., 9, 1959.

Yoshizawa, H. [1]: Some remarks on unitary representations of the free group, Osaka Math. J., 3, 1951.

Yosida, K.–Kakutani, S. [1]: Operator theoretical treatment of Markov processes and mean ergodic theorem, Ann. Math., 42, 1941.

Zaanen, A. C. [1]: Linear analysis, Amsterdam, 1953.

INDEX

Mathematics

FUNCTIONAL ANALYSIS (Second Corrected Edition), George Bachman and Lawrence Narici. Excellent treatment of subject geared toward students with background in linear algebra, advanced calculus, physics and engineering. Text covers introduction to inner-product spaces, normed, metric spaces, and topological spaces; complete orthonormal sets, the Hahn-Banach Theorem and its consequences, and many other related subjects. 1966 ed. 544pp. 6⅛ x 9¼. 0-486-40251-7

ASYMPTOTIC EXPANSIONS OF INTEGRALS, Norman Bleistein & Richard A. Handelsman. Best introduction to important field with applications in a variety of scientific disciplines. New preface. Problems. Diagrams. Tables. Bibliography. Index. 448pp. 5⅜ x 8½. 0-486-65082-0

VECTOR AND TENSOR ANALYSIS WITH APPLICATIONS, A. I. Borisenko and I. E. Tarapov. Concise introduction. Worked-out problems, solutions, exercises. 257pp. 5⅜ x 8¼. 0-486-63833-2

AN INTRODUCTION TO ORDINARY DIFFERENTIAL EQUATIONS, Earl A. Coddington. A thorough and systematic first course in elementary differential equations for undergraduates in mathematics and science, with many exercises and problems (with answers). Index. 304pp. 5⅜ x 8½. 0-486-65942-9

FOURIER SERIES AND ORTHOGONAL FUNCTIONS, Harry F. Davis. An incisive text combining theory and practical example to introduce Fourier series, orthogonal functions and applications of the Fourier method to boundary-value problems. 570 exercises. Answers and notes. 416pp. 5⅜ x 8½. 0-486-65973-9

COMPUTABILITY AND UNSOLVABILITY, Martin Davis. Classic graduate-level introduction to theory of computability, usually referred to as theory of recurrent functions. New preface and appendix. 288pp. 5⅜ x 8½. 0-486-61471-9

ASYMPTOTIC METHODS IN ANALYSIS, N. G. de Bruijn. An inexpensive, comprehensive guide to asymptotic methods—the pioneering work that teaches by explaining worked examples in detail. Index. 224pp. 5⅜ x 8½ 0-486-64221-6

APPLIED COMPLEX VARIABLES, John W. Dettman. Step-by-step coverage of fundamentals of analytic function theory—plus lucid exposition of five important applications: Potential Theory; Ordinary Differential Equations; Fourier Transforms; Laplace Transforms; Asymptotic Expansions. 66 figures. Exercises at chapter ends. 512pp. 5⅜ x 8½. 0-486-64670-X

INTRODUCTION TO LINEAR ALGEBRA AND DIFFERENTIAL EQUATIONS, John W. Dettman. Excellent text covers complex numbers, determinants, orthonormal bases, Laplace transforms, much more. Exercises with solutions. Undergraduate level. 416pp. 5⅜ x 8½. 0-486-65191-6

RIEMANN'S ZETA FUNCTION, H. M. Edwards. Superb, high-level study of landmark 1859 publication entitled "On the Number of Primes Less Than a Given Magnitude" traces developments in mathematical theory that it inspired. xiv+315pp. 5⅜ x 8½. 0-486-41740-9

CALCULUS OF VARIATIONS WITH APPLICATIONS, George M. Ewing. Applications-oriented introduction to variational theory develops insight and promotes understanding of specialized books, research papers. Suitable for advanced undergraduate/graduate students as primary, supplementary text. 352pp. 5⅜ x 8½.
0-486-64856-7

COMPLEX VARIABLES, Francis J. Flanigan. Unusual approach, delaying complex algebra till harmonic functions have been analyzed from real variable viewpoint. Includes problems with answers. 364pp. 5⅜ x 8½. 0-486-61388-7

AN INTRODUCTION TO THE CALCULUS OF VARIATIONS, Charles Fox. Graduate-level text covers variations of an integral, isoperimetrical problems, least action, special relativity, approximations, more. References. 279pp. 5⅜ x 8½.
0-486-65499-0

COUNTEREXAMPLES IN ANALYSIS, Bernard R. Gelbaum and John M. H. Olmsted. These counterexamples deal mostly with the part of analysis known as "real variables." The first half covers the real number system, and the second half encompasses higher dimensions. 1962 edition. xxiv+198pp. 5⅜ x 8½. 0-486-42875-3

CATASTROPHE THEORY FOR SCIENTISTS AND ENGINEERS, Robert Gilmore. Advanced-level treatment describes mathematics of theory grounded in the work of Poincaré, R. Thom, other mathematicians. Also important applications to problems in mathematics, physics, chemistry and engineering. 1981 edition. References. 28 tables. 397 black-and-white illustrations. xvii + 666pp. 6⅛ x 9¼.
0-486-67539-4

INTRODUCTION TO DIFFERENCE EQUATIONS, Samuel Goldberg. Exceptionally clear exposition of important discipline with applications to sociology, psychology, economics. Many illustrative examples; over 250 problems. 260pp. 5⅜ x 8½.
0-486-65084-7

NUMERICAL METHODS FOR SCIENTISTS AND ENGINEERS, Richard Hamming. Classic text stresses frequency approach in coverage of algorithms, polynomial approximation, Fourier approximation, exponential approximation, other topics. Revised and enlarged 2nd edition. 721pp. 5⅜ x 8½. 0-486-65241-6

INTRODUCTION TO NUMERICAL ANALYSIS (2nd Edition), F. B. Hildebrand. Classic, fundamental treatment covers computation, approximation, interpolation, numerical differentiation and integration, other topics. 150 new problems. 669pp. 5⅜ x 8½. 0-486-65363-3

THREE PEARLS OF NUMBER THEORY, A. Y. Khinchin. Three compelling puzzles require proof of a basic law governing the world of numbers. Challenges concern van der Waerden's theorem, the Landau-Schnirelmann hypothesis and Mann's theorem, and a solution to Waring's problem. Solutions included. 64pp. 5⅜ x 8½.
0-486-40026-3

THE PHILOSOPHY OF MATHEMATICS: AN INTRODUCTORY ESSAY, Stephan Körner. Surveys the views of Plato, Aristotle, Leibniz & Kant concerning propositions and theories of applied and pure mathematics. Introduction. Two appendices. Index. 198pp. 5⅜ x 8½. 0-486-25048-2

INTRODUCTORY REAL ANALYSIS, A.N. Kolmogorov, S. V. Fomin. Translated by Richard A. Silverman. Self-contained, evenly paced introduction to real and functional analysis. Some 350 problems. 403pp. 5⅜ x 8½. 0-486-61226-0

APPLIED ANALYSIS, Cornelius Lanczos. Classic work on analysis and design of finite processes for approximating solution of analytical problems. Algebraic equations, matrices, harmonic analysis, quadrature methods, much more. 559pp. 5⅜ x 8½. 0-486-65656-X

AN INTRODUCTION TO ALGEBRAIC STRUCTURES, Joseph Landin. Superb self-contained text covers "abstract algebra": sets and numbers, theory of groups, theory of rings, much more. Numerous well-chosen examples, exercises. 247pp. 5⅜ x 8½. 0-486-65940-2

QUALITATIVE THEORY OF DIFFERENTIAL EQUATIONS, V. V. Nemytskii and V.V. Stepanov. Classic graduate-level text by two prominent Soviet mathematicians covers classical differential equations as well as topological dynamics and ergodic theory. Bibliographies. 523pp. 5⅜ x 8½. 0-486-65954-2

THEORY OF MATRICES, Sam Perlis. Outstanding text covering rank, nonsingularity and inverses in connection with the development of canonical matrices under the relation of equivalence, and without the intervention of determinants. Includes exercises. 237pp. 5⅜ x 8½. 0-486-66810-X

INTRODUCTION TO ANALYSIS, Maxwell Rosenlicht. Unusually clear, accessible coverage of set theory, real number system, metric spaces, continuous functions, Riemann integration, multiple integrals, more. Wide range of problems. Undergraduate level. Bibliography. 254pp. 5⅜ x 8½. 0-486-65038-3

MODERN NONLINEAR EQUATIONS, Thomas L. Saaty. Emphasizes practical solution of problems; covers seven types of equations. ". . . a welcome contribution to the existing literature...."–*Math Reviews.* 490pp. 5⅜ x 8½. 0-486-64232-1

MATRICES AND LINEAR ALGEBRA, Hans Schneider and George Phillip Barker. Basic textbook covers theory of matrices and its applications to systems of linear equations and related topics such as determinants, eigenvalues and differential equations. Numerous exercises. 432pp. 5⅜ x 8½. 0-486-66014-1

LINEAR ALGEBRA, Georgi E. Shilov. Determinants, linear spaces, matrix algebras, similar topics. For advanced undergraduates, graduates. Silverman translation. 387pp. 5⅜ x 8½. 0-486-63518-X

ELEMENTS OF REAL ANALYSIS, David A. Sprecher. Classic text covers fundamental concepts, real number system, point sets, functions of a real variable, Fourier series, much more. Over 500 exercises. 352pp. 5⅜ x 8½. 0-486-65385-4

SET THEORY AND LOGIC, Robert R. Stoll. Lucid introduction to unified theory of mathematical concepts. Set theory and logic seen as tools for conceptual understanding of real number system. 496pp. 5⅜ x 8¼. 0-486-63829-4

TENSOR CALCULUS, J.L. Synge and A. Schild. Widely used introductory text covers spaces and tensors, basic operations in Riemannian space, non-Riemannian spaces, etc. 324pp. 5⅜ x 8¼. 0-486-63612-7

ORDINARY DIFFERENTIAL EQUATIONS, Morris Tenenbaum and Harry Pollard. Exhaustive survey of ordinary differential equations for undergraduates in mathematics, engineering, science. Thorough analysis of theorems. Diagrams. Bibliography. Index. 818pp. 5⅜ x 8½. 0-486-64940-7

INTEGRAL EQUATIONS, F. G. Tricomi. Authoritative, well-written treatment of extremely useful mathematical tool with wide applications. Volterra Equations, Fredholm Equations, much more. Advanced undergraduate to graduate level. Exercises. Bibliography. 238pp. 5⅜ x 8½. 0-486-64828-1

FOURIER SERIES, Georgi P. Tolstov. Translated by Richard A. Silverman. A valuable addition to the literature on the subject, moving clearly from subject to subject and theorem to theorem. 107 problems, answers. 336pp. 5⅜ x 8½. 0-486-63317-9

INTRODUCTION TO MATHEMATICAL THINKING, Friedrich Waismann. Examinations of arithmetic, geometry, and theory of integers; rational and natural numbers; complete induction; limit and point of accumulation; remarkable curves; complex and hypercomplex numbers, more. 1959 ed. 27 figures. xii+260pp. 5⅜ x 8½. 0-486-63317-9

POPULAR LECTURES ON MATHEMATICAL LOGIC, Hao Wang. Noted logician's lucid treatment of historical developments, set theory, model theory, recursion theory and constructivism, proof theory, more. 3 appendixes. Bibliography. 1981 edition. ix + 283pp. 5⅜ x 8½. 0-486-67632-3

CALCULUS OF VARIATIONS, Robert Weinstock. Basic introduction covering isoperimetric problems, theory of elasticity, quantum mechanics, electrostatics, etc. Exercises throughout. 326pp. 5⅜ x 8½. 0-486-63069-2

THE CONTINUUM: A CRITICAL EXAMINATION OF THE FOUNDATION OF ANALYSIS, Hermann Weyl. Classic of 20th-century foundational research deals with the conceptual problem posed by the continuum. 156pp. 5⅜ x 8½. 0-486-67982-9

CHALLENGING MATHEMATICAL PROBLEMS WITH ELEMENTARY SOLUTIONS, A. M. Yaglom and I. M. Yaglom. Over 170 challenging problems on probability theory, combinatorial analysis, points and lines, topology, convex polygons, many other topics. Solutions. Total of 445pp. 5⅜ x 8½. Two-vol. set. Vol. I: 0-486-65536-9 Vol. II: 0-486-65537-7